THE MISINFORMATION OF COVID-19

THE MISINFORMATION OF COVID-19

Written By:
Austin Mardon
Angela Kazmierczak
Hannah Schepian
Jake Schepian
Daniel Sperling
and Catherine Mardon

Cover Design By:
Kiran Rai

First Printing: 2020

Typeset and Cover Design by Kiran Rai & Josh Harnack

ISBN: 978-1-77369-169-5

Golden Meteorite Press
103 11919 82 St NW
Edmonton, AB T5B 2W3
www.goldenmeteoritepress.com

TABLE OF CONTENTS

PREFACE

It is unlikely Canadians thought a frenzy for toilet paper, hand sanitizer, or ground hamburger would strike in Canada. It is also improbable Canadians thought citizens would speculate the origins of a virus and how it had spread, or the very existence of one.

While the coronavirus separated us two metres apart or locked us indoors for months on end, ideas still proved more dividing; with the outpour of real and bologna news coverage coupled with personal opinions being plastered on social media, we had polarized in our beliefs about the virus. To our demise, those conflicting messages eventually manifested into emotions and behaviour.

For example, rules to distance or to wear a mask became measured according to the beliefs we held about the virus. We witnessed it firsthand at grocery stores, when distancing or wearing a mask became a warty exchange or a devout, religious practice; when online,, ideas of the virus' origins became a test for friendships or an indication of level headedness. In comment sections, when people questioned sources and the precautions taken by citizens, businesses, and health professionals, while others jabbed they were not proactive or strict enough in combating the droplet-travelling bacteria. The world was submerged in confusion, battling a cesspool of perplexing information.
What were Canadians to believe about the crisis? How

would you tell if you had bitten deeply into bologna or heard the truth? What were Canadians supposed to do anyways?

Whether we had agreed or disagreed with all of the Corona Virus Disease of 2019 (COVID-19) parameters, we all still, somehow, unify in a grounding struggle with the communications related to the virus. In our flounders of what to believe, to dismiss as old news, or to scrap altogether, the Antarctic Institute of Canada has compiled a team of writers from various disciplines, from businesspeople to becoming-doctors to aspiring writers, to dig into the misinformation of COVID-19 and dismantle it. We hope our efforts to open communication channels and to help in navigating through these times points us all true north and to solace.

Lastly, a huge thank you to Antarctic Institute of Canada's project leaders. We are fortunate that the need for clearing communications inspired Dr. Mardon and the Antarctic Institute of Canada to reform communicators across Canada to help provide information Canadians craved and needed to move forward during the pandemic, safely. We appreciate the emails, Zoom meetings, and efforts delegated to making this possible. A big thank you.

INTRODUCTION

Everything felt to have changed overnight. News outlets were swarming on updates of the Coronavirus Disease of 2019 (COVID-19), reporting threads of information from government officials and the World Health Organization (WHO), the International Health Representatives of the United Nations (UN). Reports from WHO of an outbreak in northern Italy of a novel coronavirus, meaning an unknown illness, had surfaced on February 24th, 2020.

Prior to WHO being notified by countries, China and Italy, of a rise in pneumonia deaths, cases of pneumonia had sprung only irregularly in December of 2019, beginning in China, Thailand, Japan, and eventually in Western countries, such as America and Canada (Kantis et al., 2020, p. 1). First believed as bad cases of pneumonia, they would later be identified from The Lancet, a study from China, as a novel coronavirus, located in the lower respiratory tract (Huang et al., 2020, p. 497). As the spread of the virus continued, doctors begun to suspect previous pneumonia cases to be the coronavirus. It was officially a global pandemic.

Days after warnings from WHO on February 24th, 2020, government leaders from around the globe started closing borders in fear of the virus spreading, first closing in North Korea and rippling to 50 other countries, including in Asia, Europe, South America, Africa, the Middle East, North America, and so forth. To say the least, borders known to stay open were based on no suspected threat of the virus. An article from the Pew Research Center claims 91% of countries enforced travel restrictions on travellers neither citizens or residents (Connor, 2020, para. 1). That means a big chunk of the 195 countries of the world had closed entry to outsiders. The world had stilled in its activities.

But, as though in a ripple, a thicket of fear fell rampant over Earth, as, across the globe, people locked themselves indoors out of fear of catching the bacteria or paying costly tickets. Upon surfing our news feed every morning for updates, it felt disasters happened one after the next, a year spiraling to a 2020 Antibacterial Doomsday and beyond. As buzz of the virus from Wuhan, China took hold of news platforms, the question of toilet paper came rolling in by the square for Canadians.

Catching wind of a shortage, Canadians raced to grocery stores for the last of toilet paper and also purchased irregular amounts of paper towel, face masks, and dried goods, shopping in mind of the unpredictability of upcoming months. To their dismay, previously stocked shelves bared a dusty surface from panicked shoppers buying in 5s, 10s, or by 20s, spurring yet another cycle of panicked shopping. Retailers soon enforced limits on pasta, toilet paper, meats, and canned goods so Canadians across the country had access to goods.

Whiffs of a shortage of toilet paper had even fueled urban shoppers to drive to small towns and quaint gas stations for thinly cut toilet paper, while jokes of using snow or pine combs became the norm, a way of coping. Shoppers, alarmed at the empty shelves of toilet paper, had predisposed it was the end of our chain supply. Though, according to Statistics Canada, it was the increase in demand by 250% in March that disrupted the demand chain (2020, p. 5).

As the hysteria of COVID-19 spurred splurging and solitude, for health professionals it of course meant working overtime tending to the sick or deceased or for truck drivers to work longer shifts behind the wheel; the virus demanded different duties from us all. A saddening time across the world, we depended on online communication to diffuse our uncertainties' to only find uncertainty, political head bunting, and deception.

In the coming weeks, as businesses closed or enforced safety protocols, medical research centered on prevention, what the virus was, and its origins. The origins of the virus quickly turned into a prickly conversation, looming with suspicion and pointed questions from the fearful, disheartened, a by-product occurring from anxiety for the future. Social media and news alike began surfacing videos of the unsanitary practices and questionable wild foods sold at the Wuhan Food Market, where, according to WHO, the virus is believed to originate from (World Health Organization, 2020, p. 1).

China, known for its wet markets, meaning fresh, live animal markets, became the spectacle for investigation; as the handling and mixing of animals and wild animals, vendors and buyers identified as a breeding ground for the transmission of diseases (Maron, 2020, para. 1-15).

Bats for sale, dogs packed in rusty cages, and piled high squeaky bamboo rats also called into question the eating practices embraced in Wuhan and the rest of China (Daly, 2020, para. 2). It was soon afterwards the Chinese government began cracking down on wild farms and wet markets susceptible to alarming or spreading viruses (para. 2).

Further research into China's cultural eating practices, we realized underlying assumptions about Chinese eating practices flooded the media, and was fueled by common generalizations. It was true: many had consumed bowls of boiling bat soup, but not as many bowls as social media purported (para. 2-3). It became a gap in need of informing.

An article from May in The Wallstreet Journal later discussed scapegoat Huanan Seafood Market, located in southern Wuhan, was merely a place of transmittal for the virus; the origin was leaps different than theorized (Ridley, 2020, para. 1). There was a duty to separate speculation from fact, from old information and the new, given the mounts of information emerging. That led to further questions, what is known about the virus today and what is deceptive? How do you sniff the false from the legitimate apart, or those in-between? Important to realize, but the virus bred grounds for bogus information, including scams, fake antidotes, and other realms of unaccounted issues; all of which we'll be discussing.

And yes, although misinformation issues already lurked in the background prior to the coronavirus, our friend, Misinformation, showed its face more clearly and free from restraint during the crisis. The magnitudes of reports commenting on COVID-19 conjoined with opinions and speculation from conspiracy videos only intensified misinformation's pull.

Upon surfing the internet, Canadians encountered misleading headlines, articles intentionally spreading deceptive, false information about preventative measures; then a handful of them hoaxing or grooming beliefs to a singular position; and then also some articles that strictly released research and figure-based regurgitations about the outbreak, explaining neat conclusions about molecular chains, but still reading as part-gibberish and doubted by researchers. Information overload had swayed us in a thousand directions, preventing us from coming to terms with the world's cry.

Our not knowing of how to control booming information from the internet, an information overload an encyclopedia likely couldn't contain in print form, made it difficult to differentiate the legitimate from the illegitimate or how to even go about fixing communications.

We could monitor communications more rigorously, but then the notion of infringing personal freedoms interplayed. However, if we were educated on how to handle a slew of information, that could address the issue of misinformation at its worming core. That could be our easy peasy way out. Considering the amount of time spent by the average person on social media, which is roughly 24 hours a week, it only makes sense to educate our usage and to be informed consumers of media (Hall & Li, 2020, para. 2).

Misinformation had been a slippery friend in need of boundaries for a long time. But aside from deliberate misinformation, there are cases, and many of them, of people unintentionally spreading false information. In our pondering of the errs of communication, it is easy to mix up words, misunderstand a speaker, or not select the right word for describing a situation. How much easier it must be for the focus of long-winded stories or posts to spur unintentional generalizations or ideas, which only further suggests the severity of forming opinions hastily.

In retrospect, we saw the media offered a dip into a cesspool of biases and suggested the allusion of a fully painted picture through a single post—if we weren't vigilant.

That being said, more than ever, it is important for media-consumers to seek multiple sources and to consider perspectives beyond what is offered. More than ever, educating how to engage with information is a matter of unity and safeguarding against harmful biases. Yes, how to interact with information from a given snippet of time, to evaluate assumptions, and in how to form opinions is crucial for our relationships. We are immersed in a time of combating fake information and endless generalizations, and, to win, it starts from looking within. The media needs to be disseminated and torn apart, researched beyond what a snippet of time purposes, although it may or may not be misguided. However, we realize we can only improve by sifting our ideas, not by hiding them. And that's where others can step in.

Combating misinformation is exposing lies for prevention of a global-scale bondage. And through discussing ideas and overriding personal biases, we can make ample room for improvement. There you have it: breaking topics apart into its savoury chunks, we hope to free topics laced with a nuance of lies or unknowns. It is time for our slippery friend, Misinformation, to be exposed for what it is: misleading and deceptive information.

Chapter 1:
COVID-19 History vs. Other Pandemics

First things first, a pandemic is "an epidemic occurring worldwide, or over a very wide area, crossing international boundaries and usually affecting a large number of people" (Heath, 2011, para. 1). When officials first referred to COVID-19 as a global-pandemic, and we quietly scratched our heads in private, that is what they meant. Application of the definition makes it clear that the world was ensnared in a pandemic, as we failed to contain it and saw an incline in COVID-19 cases on a regular basis in our counterparts. But, luckily, it wasn't the first pandemic the world had faced. And thankfully we weren't born during those pandemics. As we look at and compare other pandemics to COVID-19 in this chapter, people wishing they were born in another era might refrain after hearing the gory details. I mean, our odds of survival looked skimpy.

I know, history is not for everyone, but what's intriguing is our handling of COVID-19 dates to the great plagues of the past; it assuages our grappling with why we wear masks or quarantine the sick.

While it is valid that online misinformation was a different culprit to overthrow in our COVID-19 pandemic, compared to pandemics from the 1300s, it was previous pandemics that expanded our knowledge about hygiene practices and in how to contain viruses. To go even further, I would say that while misinformation has been running rampant, teeter-tottering a zillion ideas, we were, to a degree, greater equipped in containing pandemics from past episodes.

In looking at past pandemics, which occurred more often than imagined, it not only reveals the hardships in containing diseases, but also testifies of heroic, everyday citizens. We realize, our combatting of misinformation and COVID-19 are what today's generation can lend to the future and breathe life into. As history taught us about handling struggles or warring against complex ideas, we can help the future make sense of their today, in terms of containment, COVID-19, and misinformation.

In this chapter, we will admire how people who pledged war on combating infectious viruses enriched us with hard-earned knowledge. Knowledge that required costly trial-and-error runs, often resulting in their demise.

From herb infused masks to red crosses marked on doors, we will be looking at and comparing three past pandemics to COVID-19. As you read about history, which often feels a millennia away, of a dusty feel to it, you might feel as we did, that our similarities meshed us inches closer.

THE BLACK DEATH (1350)

The Black Death was a global pandemic that hit Europe from 1347 to 1352 CE. According to History (2010), the plague had arrived in the rainy October of 1347, "when 12 ships from the Black Sea docked at the Sicilian port of Messina (para. 1). Although the temptation presents itself to reason that the Black Sea inspired the plague's name, don't be fooled.

As people approached the ships, parked at the docks, they gasped as most inhabitants were dead, except for a few. The few, though alive, had black boils, the size of pears and blackberries, oozing pus uncontrollably. Upon further inspection, Sicilian authorities hurried to send away the ships, but were too late in doing so; their brief meeting led to the death of almost one-third of Europe's population, a total of twenty-five to thirty million people (para. 1).

The next five years, urban residents rushed to the countryside for safety or out of the country, to only spread the bubonic plague. Psst... A bubonic plague is characterized by enlarged or swollen lymph nodes, chills, headaches, fever, fatigue, and weakness (Stöppler, 2019, p. 1). The swollen lymph nodes, located in the groin or armpit area, became swollen and took on a black colour, hence the name The Black Plague.

As time progressed to years and a cure forlorn, citizens, in rebellion to leaders, abandoned villages and towns. It would take nearly 200 years for Europe's population to recover to original numbers (Cartwright, 2020, para. 1).

Researchers emphasize that questionable sanitation and a lack of understanding in microscopic organisms had contributed to the spread of the virus (para. 7). Pointing our finger at their sewage system feels tempting right about now, but let's refrain ourselves; they simply didn't have access to the wealth of resources we have today. Though, they, as a matter of fact, concluded proximity interplayed.

Legislation in Ragusa, Dubrovnik from 1377 required ship crews to spend forty days on-board before entering areas susceptible to transmitting diseases (Conti, 2008, p. 456). In fact, the term quarantine, to isolate, originates from the Italian words, trentino (thirty days) or quarantino (forty days) (p. 455). The legislation had first required thirty days, but changed to forty because of Biblical implications (p. 455). In the Old Testament, the Bible encouraged people to isolate the sick and burn their clothes, particularly from those with infectious leprosy (p. 455). Officials also believed that by quarantining crew members, signs of illness would present itself within those forty days. Though survival was not promised for the isolated, a quarantine ensured the disease was contained, thus preventing another tragedy.

The effectiveness of quarantining is questioned today and sometimes rebelled against. It is obvious the quarantine was an effective means of protection in the past; rather than a body of people dying, the few infected had died instead.

Travelling at sea or trading was a dreary mission for crew members, as a million things could go awry from weather conditions, but combating disease was a risk crews faced knowingly. In their sacrifice, they helped build our world. Today, circumstances are different than the 1300s, and the duration of plagues, shorter. Well, in some cases pandemics have shortened, having more insight into battling contagions from previous struggles. However, that knowledge sometimes makes not quarantining sound seductive, or even logical, but if we don't have a cure to that novel disease, it is still best to isolate like the Europeans did, until then.

THE SPANISH FLU (1918)

It has been 566 years since The Black Plague, and society is challenged by another plague. The Spanish Flu, although its origin is unknown, started in February of 1918 and ended in April of 1920. Upon our studying of The Spanish Flu of 1918, a time not long ago, humankind's vulnerability never felt more tangible. For two years, people of all ages and from around the globe, died.

Spreading through Spain, France, Great Britain, and Italy, the disease eventually infected the military during World War I, passing to Americans and other troops (Martini, 2019, p. 64). The disease, that passed across borders through carriers, caused the death of more than 50 million people. Health authorities had staged a "certain front" about the infection's etiology, the causation, although isolating the Pfeiffer's bacillus was riddled with difficulty (p. 64-65).

The first wave was not regarded as influenza by some countries, but the mutated strain of the second wave was without doubt more than a cold (p. 64). Most deaths occurred during the second wave, lasting shy of six weeks (p. 64). The first wave's symptoms included chills, a fever, and fatigue, but the second wave resembled that of a deathly horror scene; skin turned blue, nasal hemorrhages occurred, people suffocated from fluid filling the lungs, blood-streaked urine filled bowls, comas—to only die within hours or days.

The lethal flu contaminated all age groups, and had intensified once troops, refugees, and extra-working women united after World War I (p. 65). Knowing that the virus spread through respiratory droplets—from coughing, speaking, sneezing, or touching infected surfaces—officials enforced restrictions or surveillance over risky public areas (p. 65). Despite officials' attempts to plateau the curve, protocols proved futile in combatting the virus (p. 65). For example, in New York, rigorous cleaning practices happened; busing schedules changed; spraying of disinfectant filled the streets; obligatory notifying of cases became implemented; but, still it revealed not to be enough; only quarantining, isolating, and identifying suspected cases, as demonstrated by Italians, ebbed the Spanish Flu's curve (p. 65).

Intentions behind strict disinfecting and distancing was to contain the virus until a vaccine or antibiotic emerged (p. 65). People were cautious of close proximities and lengthened distancing, to a degree, but still isolating and quarantining showed most effective at fighting the virus. The plummet of those infected in Italy suggested this.

While the Spanish Flu, to a degree, deterred World War I pursuits, as environments fell prey to illnesses, the war also roadblocked the unhinging of the virus. Communication between many countries severed from the war, the discovery of a cure slowed. Perhaps if news outlets weren't targeted during the war and used for propaganda, insights about the virus would have emerged, breaking ground in its origins or sooner in expelling it.

The Spanish Flu stands as a milestone in fighting pandemics though, as new methods were applied. The major breakthrough in understanding viruses, the discovery of transmittal via respiratory droplets by German biologist Richard Pfeiffer in 1898, was executed in The Spanish Flu's proactive measures. It had led to disinfecting streets and regulating public places, setting a new approach in killing contagions at the root. However, although diminishing transmittal through disinfectant was perceived unsuccessful by some, the strictness of adherence to it is debatable (p. 65).

Today, we are still learning how to execute Pfeiffer's breakthrough in virus transmission. Although methods of expelling contagions in public settings have not exceeded the effectiveness of isolation, professionals will continue searching for the antidote or method. And, lastly, from obtaining awareness of pandemic history and of the trials in prevention, it can weaken society's resistance to proactive measures, strengthening our curbing of the virus.

SARS: Severe Acute Syndrome (2003)

Out of the three pandemics studied, the Severe Acute Syndrome (SARS) strikes the strangest, almost like a spooky mystery; it also mimics COVID-19 in some ways.

After suspected holes in coverage and a distrust in media reportage, Chinese citizens began whispering of an outbreak (Qiu et al., 2018, p. 2). And, suspicions were accurate. In November of 2002, a threatening outbreak of SARS in Guangdong, China occurred (Abdullah et al., 2003, p. 1042).

Panicked, Chinese citizens took prevention into their hands by wearing masks and spreading word of a medicinal concoction that killed the virus. Citizens of the region had begun drinking woad, a flowering plant known as "Radix Isatidis," "Asp of Jerusalem," or "Blue Dye", mixed with vinegar (Qiu et al., 2018, p. 2). Rumours of a cure spreading to other regions of China, panic shopping kicked in; stores sold out of woad and antiviral drugs in a month's time (p. 2). The researchers go on to say that because of a lack of information from news sources, misinformation bred rapidly through word-of-mouth, texting, and social media (p. 2).

Now comes the strange part: although the outbreak occurred and travelled cross-borders, the unpredictable symptoms made the outbreak difficult to scale in (Abdullah et al., 2003, p. 1042). A more comprehensive explanation: According to researchers, "the nonspecific signs and symptoms of [SARS], coupled with a relatively long incubation period and the initial absence of a reliable diagnostic test, limited the understanding of the magnitude of the outbreak (p. 1042).

Some described the symptoms of SARS as diarrhea, high fever, body aches, pneumonia, and mild respiratory symptoms, while other countries described symptoms as "muted symptoms" or even as "malaria-like symptoms" (Centers for Disease Control and Prevention (CDC), 2017, para. 3; Abdullah et al., 2003, p. 1043). SARS was perplexing for health practitioners' as SARs could easily run undetected until death (p. 1043).

Cases, believed to be SARS, were identified on a global-scale, affecting not only Asia, but North America, Europe, and South America (CDC, 2017, para. 1). Although seven hundred and seventy-seven people had died from SARS, it is believed that speedy containment of the coronavirus would have prevented more deaths, and its social and economic impact (CDC, 2017, para. 1; Qiu, 2018, para. 1).

SARS was evidently not as lethal as other pandemics, but it alludes to how misinformation sometimes spreads. Citizens felt ill-informed and worried, making them turn to the rumour-mill, social media, and texting for solutions. What is also strange is how symptoms of SARS weren't easily identifiable; it disguised as other illnesses, sometimes only unveiling itself once death reared its head.

By comparing the three plagues to COVID-19, we see recurring cycles in peoples' actions, from rebelling against authority to seeking other mediums for news of prevention. In conclusion, exploration of recurrent patterns in pandemic history might leeway to shrinking the fears of humankind, strengthening our trust in implemented proactive measures, such as isolation and quarantining. Lastly, it would be interesting to compare the volume of misinformation to other pandemics void of online communication. What would it say about misinformation, society, and communication styles? Or, has the world always been bombarded with misinformation, but through other mediums? How did the past handle misinformation and how was misinformation perceived?

Chapter 2:
The Media Responses of COVID-19

"New research shows between 80% and 90% of people consume news and entertainment for an average of almost 24 hours during a typical week." (Hall & Li, 2020, para. 2). Worldwide, people turn to the media daily to inform themselves on what is going on, and what the latest news is. There are several channels on the television that offer 24/7 news coverage, allowing people to stay informed about what's going on around the world whenever convenient for them, and countless news articles popping up every day on the internet offering easy access.

Although staying informed about what's going on around the world is important, spending too much time intaking media, especially when it is negative, could do you more harm than good. Oftentimes news will focus on bad things that are going on, and if your main source of media intake is through the television, then you'll often see the same story being covered over and over again, potentially affecting you more and more each time.

It can affect your emotional response to what is going on around you and mold your opinions on pressing matters. Ever since February 2020, it's been difficult to turn on your local news channel, flip through your newspaper or read your favourite online news sources without bumping into a mention of COVID-19. Whether it's new precautions to take, a report on the numbers of cases and deaths, or new findings about the seemingly ever-changing virus, the coverage seems to be everywhere.

Media coverage of this virus has surpassed any other previous pandemic, with a study concluding that "there were 23 times more articles in English-language print news covering the coronavirus outbreak in its first month, compared to the same time period for the Ebola epidemic in 2018" (Wahl-Jorgensen, 2020, para. 5). In addition, more people have more time to consume media coverage surrounding the virus due to the stay at home precaution measures, and the closure of a lot of workplaces bringing people to work from home. Also, businesses such as gyms, restaurants, entertainment facilities and others have been forced to shut their doors, leaving more time for people to be at home potentially watching the news.

Since the virus is continuing to evolve, scientists are still unable to say for certain how it started, the long-term effects, the best way to prevent it, etc. In saying that, new information seems to be coming out on a daily basis as they continue to research the virus and how to protect ourselves. The media offers tons of information on the coronavirus, as it has been a major focus for them over the past few months; the media has covered all of the preventative measures suggested, the suggested causes, and the transmission of the novel COVID-19.

Firstly, there have been tons of preventative measures set in place over the past few months to help "flatten the curve" of the virus and to keep everyone safe. The first case of the coronavirus in Canada was reported on January 25, 2019, of a Toronto-man who had recently travelled to Wuhan China. (Bronca, 2020, para. 1)

As the cases continued to rise at an alarming rate across the world, the World Health Organization declared the corona virus a pandemic on March 11, 2020. Following that announcement, the schools and universities shut down and moved onto an online distance learning platform, unnecessary traveling was restricted, and quarantining for 2 weeks after returning from outside of the country was a must.

It was recommended to stay at home as much as possible, only leaving when absolutely necessary, such as, for grocery shopping or work. Social distancing is required in all public spaces or you could be risking a hefty ticket, if caught by police authorities.

It is recommended by the World Health Organisation to wash your hands frequently with either soap and water or an alcohol-based hand sanitizer, to maintain a safe distance from somebody who is coughing or sneezing, to wear a mask if physical distancing is not possible, to not touch your face, and to not leave your home if you are feeling unwell. (WHO, 2020, para. 3-5). On August 1st, a bylaw was passed in Edmonton, Alberta, stating that masks were to be worn in any indoor public space. As you can see there are tons of measures that were suggested by the media to protect from the virus, and to this day they continue to change.

Some businesses are closing their doors, while others are trying to reopen; all of the changes are causing a little confusion in the general public. Unfortunately, not everybody was adhering to these measures being suggested through the media to protect themselves and others, causing rapid unnecessary spikes in cases.

The biggest issue that was being covered by the news was people, particularly younger in age, refused to social distance and went to crowded places, such as bars, beaches, houses, etc. Unfortunately, this has caused clusters of new cases around the country and resulted in potential fines for the individuals and businesses refusing to partake in social distancing measures. Since the measures are ever changing when it comes to protecting ourselves and others from the pandemic, it is extremely important to stay up to date on the latest news. Ensuring that we are staying informed is one of the first steps one can take when it comes to preventing the virus.

Secondly, a big topic being covered through the media is the assumed cause of COVID-19. The origin of this virus is very important to pinpoint and to identify, so that scientists can work to prevent a similar pandemic to reoccur in the future, and it could also be helpful in the seemingly never-ending battle against the novel coronavirus.

There have been several "causes" of this global pandemic circulating amongst the population, although a lot of them are stemming from conspiracy theories started on social media. Crazy things are being brought up and shared amongst the internet, suggesting that the virus was created in a lab by scientists to be used as a weapon or that it was created by Bill Gates in efforts to microchip people so that they can be tracked. Although both of these theories have been debunked, they were getting a lot of traction in the early months of the virus.

In this chapter, we'll be focusing on the presumed causes of the virus that were circulating through mainstream media as opposed to social media platforms such as Facebook, Twitter, Youtube, etc. The virus is said to have originated in Wuhan, China, before it spread worldwide, causing a global pandemic. Although the specifics are yet to be confirmed, it has been concluded that "by comparing the available genome sequence data for known coronavirus strains, we can firmly determine that SARS-CoV-2 originated through natural processes" (Scripps Research Institute, 2020, para. 3). With that being said, all of the genetically engineered conspiracies surrounding the virus could be finally put to rest. It was said that the virus was most likely derived from an animal and transmitted into a human (para. 1). One big theory that was circulating mainstream news is that the coronavirus stemmed from bats, and a bunch of stories came out about people in China eating these animals. Although it is possible that the virus stemmed from bats, the World Health Organization has yet to confirm and has said there has been no documentation or direct human to bat contact. According to Ben Embarek, a WHO expert, "We have some kind of missing link in the story between the origin of the virus and when it started to circulate in humans" (Gale et al., 2020, para. 10). With that being said, mainstream media is continuing to cover any evidence that is emerging regarding the origins of the virus and is waiting on a definite conclusion.

Another topic being covered amongst news sources is the transmission of the virus. It wasn't until about a month after the first findings of the virus emerged that the World Health Organization released a statement claiming that it is possible for the virus to be transmitted from human to human. Soon after, it was widespread across news outlets that the virus was predominantly transmitted through the droplets that leave our body when we cough, sneeze or even talk. This caused panic within the people, leading to some wearing gloves and masks when leaving the house; it also led to some sanitizing absolutely everything before using products and first wearing gloves when handling food, groceries, containers, mail, etc. Recently, Global News has come out and reported that the virus "may spread through the air more than we once thought" (D'Amore, para. 1, 2020). Although airborne transmission is said to be less frequent than the larger droplet transmission previously mentioned, a group of scientists warned that smaller droplets can suspend and "glide through the length of a room and infect people when inhaling" (para. 9). Again, all of the possibilities of transmission are currently still being investigated, and news outlets are keeping up to date, reporting on any new information about the virus. Due to this information being released, businesses, cities, regions and even entire states are making it mandatory to wear a mask when in public indoor spaces. Wearing a mask is to help prevent any further transmission, but people still, even after fining them, aren't complying.

Lastly, the effects of this flood of information, and coverage surrounding the coronavirus, has huge impacts on the people who consume it, especially during times when people barely left their houses; they had tons more time to sit and intake more and more of it, which only lead to more and more feelings of panic, fear, and anxiety. In addition, elderly people or those with underlying medical conditions, who are more at risk at becoming hospitalized or severely ill when contracting the virus, can become very anxious from consuming the news. Researchers argue that overexposing oneself to the media can actually do more harm than good, especially when it comes to pressing matters such as the coronavirus. The researchers said that "repeated media exposure to community crisis' can lead to increased anxiety, heightened stress responses that can lead to downstream effects on health, and misplaced health-protective and help-seeking behaviors that can overburden health care facilities and tax available resources" (Garfin et al., 2020, p. 1). The massive influx of news coverage surrounding the virus can be overwhelming and confusing to some, adding on to their pre-existing anxieties about contracting the virus. Another thing to consider is the fact that news outlets tend to focus on the negatives to enforce people to consider taking safety precautions.

In conclusion, as this virus is continuing to be investigated, researchers are continuing to find new evidence on the origin, transmission, and facts surrounding the coronavirus. They are also continuing to develop more and more measures to ensure public safety and to prevent the spread or reoccurrence of the virus. With that being said, all the uncertainty is driving people to be attached to the news to find out the latest findings and evidence. The anxiety of the unknown makes people more susceptible to believing and holding on to any information they can get their hands on. In addition, with more people staying at home, they have more chances of being exposed to more news, which only leads to more fear and anxiety. Overall, people need to try and limit their intake of media and only trust credible sources, so that they can not only protect their physical health, but also their mental well-being. The news is an amazing tool to stay informed on what's going on around the world but with everything, it needs to be consumed responsibly.

Chapter 3:
Misinformation on Social Media and the Internet

In our modern society, the average internet user spends more than "40 percent of their waking life on the internet," which adds up to more than 100 days a year spent online (Kemp, 2020, para. 11). In saying that, it is safe to assume that a number of people are gathering news and information through either the internet or through social media. The biggest social media consumers are between the ages of 18-29, and 90% of that age demographic use at least one social media site (Cooper, 2020, para. 7). While social media and the internet can be very useful tools, it can be hard to control the mass of false news and misinformation. In saying that, while internet users are exposed to a plethora of information, they can oftentimes wind up mistaking misinformation for news. In fact, according to the Shearer (2018), "more people now turn to social media for news than actual newspapers" (para. 1).

When it comes to the COVID-19, people on social media went to extremes. While the social media platforms were great to share experiences and to facilitate talking about the virus around the world, they were also making it way easier to spread misinformation; it led to some people experiencing confusion, anxiety, or even danger.

All of the misinformation and fake news spread faster and easier than the actual virus, creating what the World Health Organization has named an "infodemic" (Department of Global Communications, 2020, para. 2). They describe this phenomenon as "an overabundance of information–some accurate and some not–that makes it hard for people to find trustworthy sources and reliable guidance when they need it" (World Health Organization, 2020, para. 5). There were several different categories of misinformation being spread amongst the internet and social media, whether it was the causes of COVID-19, precautions to take, at home remedies, or speculation that the whole virus was a hoax. All in all, the content that was distributed throughout social media platforms ranged from seemingly silly remarks to promoting "home remedies" that had potential to be deadly. Although social media companies such as Facebook and Twitter said they were getting a handle on all of the misinformation being spread on their platforms, there are hundreds of posts slipping through the cracks and potentially causing harm (BBC News, 2020, para. 1-20).

The rise of technology has clearly changed the everyday lives of most people in society. Social media offers a platform to open up conversations about the virus and share experiences. Unfortunately, it can sometimes lead to more harm than good. While sources such as the World Health Organization and the Centers for Disease Control and Prevention offer reliable information on social media, other sources are not offering factual information. With that being said a lot of accounts on social media are not even prioritizing reliable information to the people and are instead focusing on topics that are more likely to get more clicks, views, likes, and shares, which oftentimes tends to be more controversial or "out there" topics. Furthermore, social media platforms such as Twitter or Facebook are "designed to show content most likely to be engaged with first, whether it be accurate or not" (Allem, 2020, para. 7). These posts then get shared amongst the internet oftentimes, although the reader didn't even realize it was false information and failed to fact check before sharing it to friends and family. Another contributor to the intake of misinformation is the anxiety people feel, which fogs their mental clarity and judgement and makes them more susceptible to fake information. Another thing that can happen is people can tweak small things in a post before sharing it, completely altering the point trying to be made. Eventually, as the posts get more and more traction, it can "mutate" or be translated across languages, making it more and more false every time (Robinson & Spring, 2020, para. 17-35).

The difference between acquiring news from social media as opposed to news outlets is that social media adds a certain closeness. During an anxious time, would you rather read an article posted by a stranger? Or, an article or post that was created or shared by a friend, family member, or someone who you follow and respect? During these times, people are filled with uncertainty and are eager to take in any information, especially when it is coming from people they know and people they are close to. In addition, when a post goes viral on social media, it tends to reappear on your feed over and over as the post gets shared and interacted with. Especially, if you are on several different social media platforms, it is very common that a viral post would come up on a few different platforms. For example, if a post goes viral on Twitter, it is oftentimes also re-shared on Instagram or Facebook, etc. Although the first time you see the post you might not be quick to believe it, seeing it over and over again on numerous different platforms could slowly alter the way you perceive the information, and you could begin to believe it as the truth (Cherry, 2020, para. 14-17).

A dangerous virus or government ploy to mass vaccinate everybody? There have been numerous conspiracy theories circulating social media in regards to the virus. On one hand, there were several posts on social media questioning the severity of the virus, and even deeming it a hoax. Due to the mass influx of pandemic content flowing in everyday, it is very easy to find an "article" or a post of some sort that aligns with whatever beliefs one may have. There are many conspiracy theories circulating online such as the viral "Plandemic" video. The video suggests that the virus was planned by the government and manipulated in a lab in efforts to enforce mass vaccinations, and that wearing a mask does more harm than good, as it "activates your own virus and can make you sicker." (Lansverk & Woodward, 2020, para. 29). This video had millions of views in a matter of days and is still able to be found on the internet, despite being "taken down" by platforms such as YouTube, Twitter, and Facebook. Most claims made in the video have since been debunked several times by doctors. As the video rose to mainstream media, a lot of people began to question their trust in the medical system, and a lot of people, in particular, anti-vaxers, were convinced that COVID-19 was a hoax. (Frenkel, 2020) One of the big issues with all of this misinformation being spread online is that it is crowding the factual evidence, making people question its severity. In Canada, Global News states that "almost half of Canadians believe at least one popular coronavirus conspiracy." (Cooper, 2020, para. 1).

Unfortunately, there is a large group of people who do not believe that the coronavirus is real, and therefore are not practicing any safety measures. This doesn't only affect them and put them at risk, but also puts everyone they come into contact with as well. For Brian Hitchens, a 52-year-old Florida man, these speculations almost cost his life. Upon reading up about the virus on social media, Hitchens had deemed the virus a "fake crisis" and said that it was being "blown out of proportion" (Brito, 2020, para. 1). Hitchens didn't wear a mask for a few days and eventually got sick four and half weeks into the pandemic (para. 3-10). After showing symptoms and gradually worsening, both him and his wife ended up in the hospital, where his wife ended up having to be put on a ventilator (para. 3-5). Hitchens made a second post addressing that he "should have worn a mask in the beginning and that he is now paying the price for it" (para. 11). All in all, with all the misinformation that is speculating the severity of the virus, more and more people could refuse to take safety precautions, which could in turn potentially cost them their lives.

Next, there is a lot of speculation over social media surrounding the cause of the coronavirus. Although most evidence leads to it being caused naturally, a simple scroll through Twitter would suggest causes that are anything but natural. Theories stated how the virus was manufactured in a lab in China as a bioweapon, that the virus was spreading through 5G waves, or even that billionaire Bill Gates created the virus to implant tracking devices in people through vaccines (Barclay, 2020, para. 1-35). These types of claims can be harmful and can impact the way people are responding to the virus. For example, in several places around the world, people are vandalizing and burning down cell towers in efforts to stop technology, while others have been discriminating against Asian people, as they believe that Asians are the cause of the pandemic (Thomas, 2020, para. 1-13).

Would you believe me if I told you that inhaling the hot air from a hairdryer would cure you from the coronavirus? How about spreading around cow urine? Crazy enough, these are both some of the home remedies circulating around the internet and social media. Tweets were circulating about how snorting cocaine and drinking, or gargling bleach would cure the virus. It led to the point where news outlets had to come out with statements advising against such practises. The World Health Organization even decided to create a TikTok account, an up and coming social media platform especially popular amongst teens, to attempt to combat all of the misinformation that is being posted on the app and potentially causing harm. One home remedy in particular, alcohol, led to over 700 deaths and 5,011 people poisoned due to methanol alcohol in Iran (Al Jazeera, 2020, para. 1-4). Another shocking story was of a couple in Phoenix, who after reading a tweet from US President, Donald Trump, about the drug "chloroquine" as being a treatment against Covid-19, decided to search their cabinets to find a fish bowl additive containing the drug. They mixed the additive with liquid to drink, and within 20 minutes fell ill, to the point where the husband couldn't be revived (BBC News, 2020, para. 12-14). It is safe to say that we shouldn't believe everything we read on social media and should always do more research before ingesting cleaning products.

In conclusion, the misinformation surrounding the novel coronavirus on social media platforms trumps anything we have seen in the past surrounding illnesses. Believing this information or practicing some of the activities that are suggested in these posts could be harmful or even lethal not only to the people who are participating, but as well as for everyone around them. Although large platforms are starting to get a better grasp on monitoring the information surrounding COVID-19 that is being spread on their sites, it is for certain that some things continue to fall through the cracks. There are multiple reasons on how misinformation goes viral, and how it is believed by an alarming amount of people. Whether it be forgetting to stop and factcheck the sources, getting this information from somebody you trust, or the anxiety of the pandemic getting the best of them, believing false information could potentially worsen the virus itself. All in all, there is a mass influx of information regarding COVID-19 circulating social media platforms, and it needs to be analyzed appropriately to ensure the safety of the reader and those around them.

Chapter 4:
COVID-19 Phishing and Scams

To add to the pile of pandemic misinformation, we have COVID-19-inspired phishing and scams to guard from. While media-related content is an expected transgressor for misinformation, the rarity of consumer-fraud makes it easily undetected and full of sticky traps. Yelp.

Forgetting about the usual offenders—misleading content, fabricated news, false connections, clickbait, and so forth, we will divert into the mastermind of a con artist during COVID-19. As we reveal the tactics they employ, we will, as a result, safely harness against consumer-fraud, another form of pesky misinformation. That being said, as you suspected, this chapter focuses on how to recognize a COVID-19 ploy and what to do or avoid doing.

Background Scoop.

Phishing and scams are a form of deceptive misinformation that range in its delivery. The Merriam-Webster dictionary (2020) defines phishing fittingly, "a scam by which an Internet user is duped (as by a deceptive email message) into revealing personal or confidential information which the scammer can use illicitly (para. 1). While scams are more encompassing and define as "to deceive and defraud

(someone) (Merriam-Webster Dictionary, 2020, para. 3).
A scam may include overpricing products or obtaining
personal information in an imposed friendship. But, before
we dig into the types of scams, when did fraudulent
acts begin?

Fraud dates to the archaic time of 300 BC (Beattie, 2019,
para. 1). It is likely not the first fraudulent act, but the first
ever recorded one. A Greek merchant, Hegestratos, had
taken a "bottomry" insurance policy, a loan, for his cargo
ship and promised to pay once he sold his corn (para. 1).
Instead of settling through corn sales or attempting to repay
his loaner, Hegestratos tried sinking the ship, as a ploy,
but accidentally drowned himself and his crew (para. 2).
Today, similar heinous acts happen, but let's hone in on
specifically COVID-19 gimmicks, as it is relevant, a present-
day concern.

As far as to date, cons are disguising as health officials or
suited business people, addressing the needs or demands of
the public. From a sales point-of-view, they have identified
the angle for catching the well-meaning and taken off-
guard. It's an "intimate" breed of bogus deception to
stomp on, ruthlessly. Furthermore, the angle deployed in
COVID-19 consumer-fraud acts is also strategic.
But don't scams squeak the same squeak, you might
ask? Yes and no. While scams and phishing akin to past
dupes, COVID-19 lays a sneakier trap, as people roused in
emotional confusion and sorrow are targeted. To start, let's
look closely at COVID-19 phishing.

COVID-19 PHISHING

Cybercriminals have turned COVID-19 into an opportunity
to exploit. Criminals are reported to be sending mass
emails, claiming to be a legitimate source offering health
services or providing information about the crisis (Norton,

2020, para. 2). Once a person complies to the scheme, two things occur: malware downloads onto the owner's device, or upon a victim purchasing their product or services, credit card account details forward to the cons.

Falling prey to malware or retrieving our personal information is not difficult either. In fact, according to Kelly (2020), "Hackers know that over 90% of data breaches are the result of human error" (para. 5). Errors run as simple as accidently opening an attachment or embedded link that claims to detail COVID-19 statistics, entering emails and passwords to a fake Zoom website, or is as simple as purchasing their limited COVID-19 "cure" (Norton, 2020, para. 3).

For those wondering more about a malware attack, after the virus downloads onto the owner's desktop, the thief controls the computer, observes keystrokes, or accesses information saved on the computer for personal gain (para. 4).

It gets even trickier. To get to the point of receiving a user's information, cyberthieves act as faux bosses or health officials; they'll address or close emails with business lingo, using resonating points of reference; or provide links to a zoom meeting, pretending to be a tender boss or previous business-partner from work (Norton, 2020, para. 10; Kelly, 2020, para. 4).

To give an idea of the phony bologna to expect, here are examples of emails floating around:

- Updates from the U.S. Center for Disease Control (CDC), listing people infected with the coronavirus;
- Health advice from medical experts from Wuhan, China, offering health advice; and
- Work place emails, implementing a new policy in regards

to a COVID-19 outbreak (Norton, 2020, para. 7-10). The best way to guard against a cyber-attack is to educate the workplace, stay connected and converse with employees, discuss with others about any "weird" emails, and keep a keen eye for the sender's address (Kelly, 2020, para. 7-9). A rule of thumb, if someone sends an unexpected email detailing important information, don't open it. Question it first, look further; hover the cursor over the email address and investigate the address details. Is it an official email? Was I expecting this person to contact me? Also, pay attention to the dialogue and how it makes you feel. Cons induce loneliness, haste, burden victims with guilt, or sometimes leave a trail of spelling mistakes or glaring errors (Norris et. al., 2019, p. 231; Norton, 2020, para. 20). When receiving communication from new contacts, it is better to question than to absentmindedly forward personal information or money. And, statistics testify to it. During the first two weeks of April 2020 alone, Google reported blocking at least 18 million phishing emails related to COVID-19 (Google Cloud, 2020, para. 4).

In this sensitive and confusing time being exploited by criminals, keeping guard and digging deeper can prevent the headaches of canceling credit cards or contacting officials about a suspected case of fraud; it can keep a person from crippling debt or financial instability.

To dig deeper, all it takes is researching for official details, contacting officials to ask if they contacted you, or hovering the cursor over an email address. Be intentional in taking your time while communicating to a new contact, refrain from hasty decisions, and delete any suspicious emails. Cybercriminals may only snag a person once, but, as a matter of fact, after falling once, victims should expect two or three more phone calls requesting money (Canadian Anti-Fraud Centre, 2020, para. 12).

SCAMS

According to O'Brien (2020), "Canadians have lost more than $1.2 million in [April] to scammers taking advantage of the COVID-19 pandemic (para. 1). The Canadian Fraud Centre believes the pandemic is being used as a distraction or cover for malware attacks (para. 3).

The Canadian Fraud Centre website also illuminates some astounding statistics. From March 6th, 2020, to July 13th, 2020, Canadians have reported 2,007 cases of COVID-19 fraud, 1,067 victims of COVID-19 fraud, and a loss of $4.59 million dollars to COVID-19 fraud (Canadian Fraud Centre, 2020, p. 1).

In the grand scheme of the scamming world, non-COVID-19-related scams actually pose a nastier threat. In fact, numbers quadruple. The Canadian Fraud Centre states that as of June 30th, 2020, Canadians reported 28, 206 accounts of fraud, 13, 032 victims of fraud, and a loss of $51 million to fraud (p. 1). Compared to last year's reports, numbers suggest we have reached half of last year's totals. Not until the end of 2020, will we be able to judge, to an extent, if Canadians are catching on or if scammers are losing their potency. Worth noting, the above numbers reflect reported cases, which may not capture the volume of scams or the entirety of money lost by Canadians. It does indicate, however, that scamming is a relevant issue today. For the record, whether or not we fall for COVID-19-related snares, it is beneficial to understand the techniques scammers deploy.

Moving on, another notorious thorn for scams involves the Canadian Emergency Response Benefit (CERB). Countless news outlets, fraud alert organizations, and COVID-19 info sites flag an alert on CERB-related scams. Some describe the scam occurring via text message or phone call, through email, or on fake government websites (Canadian Anti-Fraud Centre, 2020, para. 1). Cons are reported to be offering assistance in applications for CERB, sending infected links for "CERB", or accusing victims for fraud in loan applications (para. 1).

The newest angle for CERB scams also includes applying for the loan using a fake identity or someone else's information (para. 1). Other victims report strange phone calls, bogus online ads, fraudulent charities, COVID-19 "protective" air filters, and low-quality, expired, or dangerous products, such as, disinfectant or hand sanitizer, for sale (para. 1).

You might have noticed a vicious cycle at this point. While the angle varies in phishing and scam ploys, the same cycle occurs: a person of a supposed position, who really is a disguised criminal, pries for victims' personal information or requests money, regardless of the repercussions or distress it causes for victims.

Scammers can be ruthless and sometimes go as far as accusing victims of "wrongful acts," including, masturbating in front of a computer or committing fraud in loan applications, so they succumb to paying or avoid contacting the police (Canadian Anti-Fraud Centre, 2020, para. 5). It appears it is crucial we guard ourselves through education or anti-virus programs, regardless of whether or not it is a pandemic.

Final Findings

It is commonly believed that scams target the elderly or those who speak English as a second language. However, in a study by Norris and after reading common Canadian Anti-Fraud Centre fraud-scenarios, fraudulent acts target various groups of people, depending on the scheme (Norris et. al., 2019, p. 242). Norris' et. al., (2019) study found the following:

The majority of evidence and subsequent beliefs we have regarding the psychological factors associated with vulnerability to online fraud are at best anecdotal and at worst in danger of creating misleading myths (e.g. older people are 'easy' targets). Policies designed to limit the extent and impact of fraud should clearly recognise the universal nature of compliance and that no one demographic is necessarily more or less vulnerable (Button et al. 2016). (p. 242)

Unlike the position of countless news stories or blog posts, Norton et. al., challenges thinking of the elderly as "scam-magnets," replacing it with a position in-tune with reality that lends to awareness about consumer-fraud. Succumbing to stereotypes, we discourage seeking education about the signs of fraud or implementing proactive measures. The truth is that we are all susceptible to fraud, but by creating awareness and educating ourselves we can harness against it (p. 242).

Scammers are milking on the fears of the public left and right, during the pandemic, and there are several scams and phishing attacks looming about. In this chapter, we had a taste of COVID-19 gimmicks and investigated the signs of a cyberattack, but if interested in learning more and familiarizing with other tactics or scams, I recommend visiting the Canadian Anti-Fraud Centre site for detailed information. This site, among other official government sites, gives a thorough breakdown of the tricks cons repeatedly do or are recently doing to Canadians. While we explored the scamming world and detailed some COVID-19' attacks occurring, it is also worth knowing how different communities are targeted or how young people fall prey.

Some may never encounter a scam, but the odd occurrence can have crippling repercussions to our financial standing. And although education is key in dodging scam artists, having an IT at the workplace, investing in an antivirus program, or facilitating conversations amongst colleagues about funky emails or text messages aids in prevention too; it can provide the assurance or be the safety net craved by employers or colleagues.

It can be difficult, embarrassing to talk about being scammed by a con. But after considering the manipulation, the severity of language in phishing emails, and the ease in making human errors, falling is easier than believed. Not to worry though, helpful sources are available to the public and can be accessed for free. We highly recommend giving them a peak and taking the proactive steps in protecting yourself and others from cybercrime. En garde!

CHAPTER 5:
TELECONFERENCING AND MISCOMMUNICATION

Whether you're a student, an employee working from home, or a family member catching up with friends or family, it seems as though everyone has downloaded Zoom at some point during this pandemic. 30 years ago, if someone had told you that a computer program could allow you to be in the same virtual room with up to 500 other people, most people would have probably shaken their heads. We can now visit with grandma, finish our school paper, and submit our boss's work project all without pants on from the comfort of our own home. Technology has truly changed the way we communicate and has revolutionized the workforce. The pandemic has given us a unique opportunity to put this amazing technology into practice. Being forced to stay at home means being forced to put these programs to the test. Lectures became Vodcasts, meetings became Zoom calls, and going out for dinner became Uber Eats. More than ever, grandchildren were being called upon by their grandparents to serve as a makeshift IT support, and people were forced to adopt these new ways of interacting with the world around us. What could possibly go wrong?

What Is Communication?

There's an old saying that "talk is cheap". The reality is that this couldn't be farther from the truth. Talk is extremely valuable. From the beginning of time, talking has enabled mankind to establish its dominance in the animal kingdom and become the intelligent species that we are today. To expand, communication, as defined by Merriam Webster (n.d.), is simply "the exchange of information," but it, more largely, has allowed us to share joy, work as a team, and avoid danger (para. 1).

When reflecting on the implications of communication, it has the power to inspire or to crush a spirit; to teach or to misinform. More closely, messages can be transmitted out loud or without saying a word at all; it is a complex and hard to understand phenomenon that people have devoted their lives to studying. In fact, it is said that most of communication is non-verbal, and developing strong communication skills is crucial to having any kind of success in life (Yaffe, 2011, p. 2). It's clear that communication is very important, so what does COVID-19 have to do with all of this?

How Has COVID-19 Changed Communication?

With the COVID-19 pandemic, one thing that has undeniably suffered is our communication with one another. Most in person communication has shifted to phone calls, video chats, and emails. While working from home, popping into your boss's office to ask him or her a question is no longer viable. This question will instead have to be asked in an email. You hope that you come off sounding intelligent, but is it too wordy? Are they really going to read all of that? Will they understand the tone? So much is lost in our translation to text, and it's rare that you're going to come off the way you intended. So, maybe it's best to save the questions for the weekly meeting instead!

Although video conferencing is a revolutionary technology, it still limits our scope of interaction. The ability to read body language, facial expressions, and emotions are severely limited over a screen. You'd be surprised how much information the human brain takes in while talking to someone. In our mind, we notice the person's clothes, their posture, and shifts in volume in their voice. All these things are compromised over a video chat. You're limited to only seeing the upper half of a person, and the resolution of the video feed is often questionable at best. This leaves a lot of guesswork for our brains. Simple things like the angle of the camera and microphone volume can potentially affect the way we're coming across, or are projected, to our audience.

Another fun aspect of a Zoom call is the fight for whose turn it is to speak. If you've ever been in a Zoom call, you know exactly what we mean. Any conversation involves people talking over each other trying to get that coveted green box. By the way, the green box is a fickle being and can have a mind of its own; it may shift to someone who wasn't even talking, or take it's sweet time when changing from speaker to speaker.

There's also the aspect of being able to see yourself. A video feed of our own face can potentially make us overly self-conscious of what we're saying and how we look saying it. No matter how much you like the way you look, having a picture of yourself constantly in your field of view can be daunting.

One thing that's totally lost in this pandemic is the potential for spontaneous conversations. No more can employees meet at the water cooler to discuss the weather or their thoughts on last night's game. We mean, even those conversations about last night's "The Bachelor" are gone. Conversations in the workplace are now scheduled and structured. There is a mutually agreed upon time to start and finish social interactions.

Although being able to avoid your coworker's stories about their ex boyfriend's Instagram picture seems desirable, this need to schedule interaction can be harmful for some people. It can make it tough for people to seek out connections and can potentially lead some people being socially deprived. There's all kinds of negative things correlated with this, such as, depression, anxiety, and a general feeling of being unwell (Wickham et al., 2014, p. 1). This is most noticed by extroverts who need adequate social interaction to feel normal. Our ability to build relationships and make new friends is severely limited by these new communication methods.

Many people think that these new ways of doing things might stick around, even after the COVID-19 pandemic is over. Why would a company pay for an office when they can get people to do the same jobs from the comfort of their own homes? Why would a university have 3 classes of 500 people when they could have one online class of 1500? These thoughts make sense to a lot of people. Although there's no way to determine whether these technological shifts are positive or negative, we know for sure that intimate, enriching communication will suffer.

ZOOM BOMBING AND OTHER SECURITY ISSUES

Technology doesn't come without its issues, however. As with most things in life, the good never seems to come without the bad. Despite Zoom's ability to connect people around the globe, it suffered from major security issues early on in the pandemic (Lorenz et al., 2020, para. 3). This left many companies, schools, and families confused when things started going awry. The newfound phenomenon of 'Zoom bombing' took over the internet as pranksters and cyber criminals found their way into meetings where they didn't belong (para. 2). These "Zoom bombers" would then interrupt meetings or lectures by posting lewd, obscene, or even racist imagery (para. 1-5). Hundreds of YouTube videos made about "Zoom college class pranks" have been going viral and garnering millions of views. This created major public relation problems and raised a ton of controversy over Zoom's security. Imagine being on a call with Grandma and someone pops in and puts on some smut! The problem became so prominent that the United States of America's FBI issued an official statement warning people about these deviants (WKOW Television, 2020, para. 2). In response to these events, many workplaces and schools stopped using Zoom and switched to other services, such as, Google Hangouts. Zoom has since implemented the use of passwords, end to end encryption, and increased host controls in order to combat the problems that they were having (para. 10).

FUTURE EFFECTS OF VIDEO CHAT

When we think of the future effects of video chat instead of in-person communication, the first thing that often comes to mind is the effects on school children. If the children are unable to return to school, it most likely won't be their academic learning that will suffer the most: it will be their social learning. Grade school years are crucial periods in children's lives as they learn to navigate social groups and relationships. When kids are stuck in their parent's house doing school in front of a computer, they miss out on so many collaborative skills that are developed in person.

It is not just the children who will be affected. As mentioned above, the lack of social interaction in the workplace can be harmful for adults too (Wickham et al., 2014, p. 1). Social interaction plays a huge part in people's well-being (p. 1).

HOW TO SOLVE THESE ISSUES

The clear trend in this pandemic is that there are often no right answers. The best we can do is to make the most of what we have. If meeting in person is a viable option, then do it! With that said, it's best to wear masks, socially distance, and do everything you can to stay safe.

If you need to meet over video chat, make the best of that situation. Having a good microphone and webcam can go a long way into making sure that people are picking up what you're putting down. Being able to better display your facial expressions and tone of voice will ensure that people better understand your message. If all you can do is email, make the most of the email! Be clear, concise, and make sure you're conveying what you're trying to convey.

As the pandemic evolves and hopefully comes to an end, miscommunication involved with teleconferencing and videoconferencing will hopefully be a thing of the past. That being said, this could potentially be the new norm. Companies that were once in an office building could now operate out of people's homes. More and more schools could be shifting towards online learning. If this turns out to be the case, it is important to keep in mind the downsides of using these methods of communication. Remember how important it is to be clear in your communication. Invest in a good webcam and microphone and make the most of the situation. Always remember to stay positive, wash your hands, and put your health and safety first.

CHAPTER 6:

MISINFORMATION IN THE POLITICAL SCENE

Your mind might catapult to Trump when thinking about misinformation in the political scene, but, believe it or not, we've uncovered more than speculative discussions about injecting disinfectant. Or, on another note, you may be thinking about whether or not misinformation exists in the political scene. Not to inspire suspicious ills toward our leaders or others, but it is good we ask ourselves these questions about misinformation. It is good to construct informed opinions and to evaluate proposed opinions— why not be a free, seasoned thinker?

For the record, we will not scrutinize past accounts of misinformation in the political arena, or discuss if it has or has not previously occurred, since we will be focusing on political misinformation' hiccups during COVID-19: how it led to Russian Intelligence writing 150 propaganda articles, what fake information politicians' relayed to the world, and, of course, the temporary suspension of Donald Trump Jr.'s Twitter account from a misleading conspiracy post on the platform (Tucker, 2020, para. 4; Brewster, 2020, para. 1).

It is important to note that in our discussions, if possible, a leaning to Conservative or Liberal views is not a basis in unearthing any intentions or cases of misinformation.

How will we draw our conclusions then? We will analyze the original source, expose what was misleading, and then include some rebuttals on the data collection; it's our attempt to rid of the political biases often attached to misinformation' accounts. In other words, political affiliation has a way of flavouring perspectives, sometimes hindering the ability to distinguish between logic and human ego. But, that is not our goal. I repeat, it's not our goal. Really.

But before cracking into the eggshells of political discussions, let's refresh ourselves of the three themes of misinformation that, according to Ricard and Medeiros (2020), boil down the essence of misinformation discussions: 1) "pseudo-scientific [suggestions] about symptoms, risks, and cures; 2) discussions of prevention and control measures adopted by countries and their collateral effects; and, lastly, 3) focus on attacking or promoting decision-makers or public figures about social isolation measures" (p. 2). In summary, researchers concluded that pandemic' misinformation exhausted those three themes, and, as we witnessed too, politicians focused on those talking points by either warning against them or actually promoting those 'against the norm' recommendations for varying reasons, such as economic stability or from belief in misleading messages (p. 3). Not to say all politicians were spreaders of misinformation by discussing those three topics, but it signals a warning for possibly sticky but critical information; we like to think of it as a snot rag susceptible to more blowing into.

While the research article drawn from, by Ricard and Medeiros, does not indicate a distinction between misinformation and disinformation until later, it explains misinformation embodies, for the most part, the following: "a minimizing of the severity of the disease, a discrediting of social isolation measures to mitigate the disease's spread, and an increasing of distrust in public data" (p. 2). Largely, as reflected from the above findings, conflict resides in prevention and in areas of major decision-making.

In the grand scheme of COVID-19, why the mass confusion in topics of life and death though? Were these simply angles that stirred the most clicks? Other questions follow too. Should a politician vocalize health recommendations? Why weren't health officials in the forefront? Why propose recommendations if they aren't 100% proven?

In the midst of a crisis, I'd hope experts take in all the information and recycle through the good and bad, providing a singular, clear direction for the government to communicate. Of course easier said than done, yet it is what's needed and then some. We acknowledge the multitude of unknowns during COVID-19 and the public's need for assurance, but in a position of leadership, as most would agree, it is best to reinforce what is known to be supported practices than to lead others into still-questionable practices.

Interestingly, numerous articles in the western world solely targeted Trump for misleading information. However, as illuminated in this study from Brazil by Ricard and Medeiros, blame falls heavily onto the shoulders of their Brazilian President, Jair Bolsonaro, for causing an influx, if not a "vector" of, the misleading content circulating on social media (p. 3).

What fueled the blaming? In live stream videos, President Bolsonaro compared the coronavirus to a gentle flu, stated 90% of the infected wouldn't experience symptoms, said "armored glass protected against the virus entering a space", and urged application of hydroxychloroquine sulfate promptly in severe coronavirus cases (p. 4). Unfortunately, a newly realized "Dr. Bolsonaro" only spread speculation, applied information out of context, and fudged information, eventually terminating social isolation measures through his misleading recommendations and mind boggling "studies" into the coronavirus (p. 4).

Our criticism of the Bolsonaro' account, in gentler terms, lies in how he took on duties beyond his role as acting president. The position of president is already a hefty enough ordeal from the looks of it. What politician or president doesn't look 10 years older after office? But most importantly, why take on more duties, especially not equipped in?

Another question, where hid the medical community? The medical community needs to speak directly to the public or, at least, speak through the president. How else can accountability be established without first the scrutiny of health practitioners?

Now onto Bolsonaro's hydroxychloroquine treatment. Although it may not be of popular opinion, there's no harm in scientists researching hydroxychloroquine sulfate or other antidotes, but it would be best practice to hold off on recommending medicines for an entire body of people or prescribing doses without approval yet from professionals. We mean, even doctors drown in trouble or face suspension after prescribing wrong or harmful medication to a patient. A key to take away, patients vary on a case to case basis.

In hindsight, the coronavirus, in many ways, became politically charged and a test of his or her leadership skills. Bolsonaro maybe was well-meaning in his attempts to give people hope or certainty, but to end social isolation measures? What was the reasoning behind that?

Fortunately, as a result of Bolsonaro and other politicians jumping off into the deep end, countless articles later communicated the need for open, frequent communication with the science community during and outside the pandemic, as it would ward off misinformation. In fact, according to Banks (2020), "Dr. Caulfield believes that scientists and public health experts must become more actively engaged in public conversation about COVID-19 and push back on misinformation (para. 24).

That proposition—more communication with the science community—seems like a timely resolution for the confusing messages surrounding the virus. Perhaps blasting supported recommendations on speakers or on social media would sooner replace the pull of misleading YouTube videos or pseudo articles. Just an idea. Or, maybe it would put a face to our professionals, making them less elitist or deemed unsuitable for objectification in suspicious conspiracy videos (para. 22).

Let's segue to Donald Trump now. Trump has been ridiculed for touting use of hydroxychloroquine sulfate to the public. Unlike Bolsonaro, Trump encouraged Americans to discuss usage of it, as a temporary COVID-19 preventive, with their health practitioner, as it fought "a broad category of illnesses known as coronaviruses, [but] not the strain that causes COVID-19" (Facher, 2020, para. 15).

Although Trump never encouraged Americans to self-administer, some later ordered on Amazon "chloroquine phosphate," an antiparasitic used in fish tanks, to only die of poisoning (para. 6). For the record, Trump stated in a May press conference that he was using hydroxychloroquine sulfate and azithromycin (Z-Pak) for prevention, as doctors disclosed positive anecdotes in letters, addressed to him, about the effectiveness of the drug on patients (CNBC Television, 2020).

It was however terribly misleading of Trump to say,

"hydroxychloroquine alone doesn't carry side effects" or "you aren't going to get sick and die [from it]" (CNBC Television).

Some wonky claims to make as president. Perhaps in Trump's case he wouldn't get sick or experience side effects, or maybe doctors had promised those misleading claims in the letters, but, still, that decision should be reserved for family doctors.

Trump though, from the sounds of it, seemed fixed on staying ahead of COVID-19 through his purchasing of staggering amounts, roughly 31 million pills, of hydroxychloroquine for temporary prevention and unpromising COVID-19 cases (CNBC Television, 2020; Facher, 2020, para. 20). Despite the unknown effectiveness of the medicine against COVID-19, at that time, he fortunately never substituted drugs for social isolation, at least not on live television (Brewster, 2020, para. 6). Yes, unfortunately, there's more to come.

Eventually, in combat of any other misleading recommendations from those in the presidential sphere, the president's son, Donald Trump Jr., was "locked out" from Twitter after posting a misleading YouTube video about hydroxychloroquine and other COVID-19 claims (para. 1). Trump too shared the YouTube video but Twitter removed it (para. 6).

Crackdown on misleading information soon intensified after yet another misleading discussion occurred from the president at the White House, when heat and light or the injection of bleach were claimed to potentially kill COVID-19 (BBC News, 2020, para. 1-4). The misleading "insights" from President Trump's advisors supposedly encouraged a crisis in Iran as social media posts hoaxed citizens into drinking bleach or alcohol for prevention of the virus (Al-Arshani, 2020, para. 8). Iranian officials, quick to dismiss the claims, were unfortunately drowned out in the purchasing of bottles of bootlegged alcohol, containing deadly methanol (para. 1). That was beyond Trump's control. Roughly over 700 Iranians died from the consumption of alcohol or bleach, after reading misleading social media posts about the alcoholic cure (para. 1).

While Trump said UV light and disinfectant were still being looked into, not to self-medicate, his sarcasm in proposing light or bleach be injected inside the body was taken literally by some (para. 10). It wasn't a suitable time for sarcasm, considering the seriousness of the pandemic and its eventual turn of events in Iran. To also add, not every country on the grid is familiar with sarcasm or the same humour.

Yet, aside from politicians recommending harmful advice, another issue surfaced: the possibility of Russian Intelligence spreading misleading articles, 150 of them (Tucker, 2020, para 4). The articles were believed to target American and Western audiences (para. 1). So, what happened? Great question. As far as we know, the U.S. government officials caught two senior Russian Intelligence members using English-language websites to spread disinformation about the coronavirus (para. 1). A CTV News article by Tucker states the following:

> The sites promot[ed] their narratives in a sophisticated but insidious effort that U.S. officials liken[ed] to money laundering, where stories in well-written English—and often with pro-Russian sentiment and anti-U.S. sentiment—[were] cycled through other news sources to conceal their origin and enhance the legitimacy of the information. (para. 10)

Well, someone definitely made the naughty list this year. Essentially, Russian Intelligence had exploited the pandemic to spread mass confusion. Although the article does not dig more deeply into the motives of Russian Intelligence, or whether they were inspired from the upcoming American election in November, they sought to "advance false narratives and cause confusion" (para. 8). To add, Russians were also allegedly trying to steal vaccine information from Canada, the United States, and the United Kingdom (Aiello, 2020, para. 1). Oh, Russia.

When would the conflict ease or halt completely in the world. Nonetheless, after learning a sum of the confusion was brought on by Russian spies, we hope it assuages some festering questions.

But on the other hand, after thinking about it, sticking all the blame to Russian spies wouldn't be fair either, as other factors contributed too. We mean, misinformation had trickled into some political meetings, bringing about even more confusion for Canadians and those abroad through online communications.

Without a doubt, the slippery, misleading recommendations from politicians caused communication chaos and division amongst citizens. Generally speaking, communication turns messy when multiple voices comment at large on an issue with their differing recommendations or propositions.

In conclusion, when leaders or those in authority spread false information, it not only affects how they are perceived, but tarnishes how future leaders are seen. Apart from politicians' reputations or striving to maintain their influence in leadership, citizens, in response, can wrongfully apply or further spread suggested claims to not only neighbours, but abroad; we saw this in Iran. What our leaders say is important. To add some food for thought, what is also not said is equally important.

Chapter 7:

Misinformation and the Rise and Fall of Businesses

The Canadian government's weathering how to tackle and scale-in on our recession, given the unpredictability of the pandemic in months ahead (Statistics Canada, 2020, para. 1). In our country and abroad, businesses have been forced to somehow adapt to the pandemic or face closure. Whilst providing supplies and economic relief from CERB or other grants were of foremost importance to leaders and business opportunists, the question of Canada's economic stability rolled in loudly. How to return to economic health post the pandemic?

According to Statistics Canada (2020), Canada's economy had been slowing down before the pandemic, as Canada's gross domestic product was unchanging in February (as cited in Evans, 2020, para. 1-2). An article by CBC suggested teacher and railway protests contributed to those stagnant numbers (para. 4). Data also highlighted how preceding February, only minuscule gains occurred in the next three months, followed by the latent introduction of the virus to the workplace (para. 3). Despite Canada's financial stagnancy and other worldly complications simmering on high, an importance rested on ridding of the virus and containing it (McKinsey & Company, 2020, para. 2). In another skin, if we hadn't emphasized containment, what could have happened health-wise, but also financially in the long run?

Upon government and health officials enacting safety protocols for citizens, health practitioners, and business owners to abide by, it resulted in drastic changes in day to day production. In Alberta, the vast majority of businesses implemented physical distancing by 2 metres (at least 2 arms lengths), disinfected surfaces, installed face guards, underwent regular questionnaire and testing, and eventually wore face masks (Government of Canada, 2020, para. 1).

In also came the question of supplies and if there were enough in stocks for the entire country, especially for those working the front-lines of healthcare. There are currently mixed beliefs in sending gear abroad. In the same vein, could we afford assisting other countries in dire need of masks, ventilators, and supplies too?

I am certain you already know of this though. Canada, the great humanitarian, sent 16 tonnes of protective equipment to China, including "clothing, face shields, masks, goggles and gloves" (MacCharles, 2020, para. 10-13). According to MacCharles, "Ottawa has defended its donation to China since it was publicly announced February 9th, saying WHO and many countries were trying to help China contain the outbreak" (para. 19).

Later, officials announced the sent products from Canada verged expiry, and regardless if they gave or did not give to China a portion of the national stockpile, they would still need to acquire or produce protective gear anyways (para. 10-21). The liberal government also explained that products would be delegated from a provincial-basis, not from the federal stockpile (para. 20).

During this time in Canada, essential businesses continued operation, except with the incorporation of safety measures. That translated into additional costs and fees, and more elbow grease to continue operating as a safe, essential business. As a side note, benefits were available to businesses to assist during the pandemic, if they met the outlined criteria. Anyway, what is considered as an essential business in Alberta? To name a few or mostly all of them: Food and supply retail stores, grocery stores, medical facilities, public security, transportation services, government and public administration, liquor and cannabis stores, financial services, water operations, energy sectors, agricultural jobs, and industrial or oil facilities (Government of Alberta, 2020, chart 1).

In response to safety measures, concerned businessowners' questioned if the virus was of enough scale or potency to halt daily operations. Many of them wondered, how could we provide the safety needed without compromising financial affairs or causing long-term debt? We are open to your ideas financial wizards. While safety is a priority for our nation, minimizing other risks susceptible to damage is lucrative to our future well-being. People were also curious if it was impossible to contain the virus without the implemented measures, namely the restrictions on essential and non-essential businesses? As discussed earlier, although we could always improve our methods, the current implemented measures protected us and others from spreading the virus, first and foremost.

Moving on, not until June 12th, 2020, non-essential businesses finally reopened, but under adherence to "general and sector-specific guidance" (para. 1). Although opening presented the risk of furthering the spread of COVID-19, non-essential businesses, including, hair salons, child cares, non-essential health services, restaurants, cafes, bars, clothing or merchandise retail stores, and wellness services, were granted to reopen but under the conditions laid out by local governance (chart 2). The businesses in the entertainment, major event, and nightclub sector continued to struggle, as they remained restricted entities (chart 3).

While businesses, meeting a certain criteria, received financial support from the federal government, others had no other choice than to close, seeing they pocketed no incoming income substantial enough to keep doors open; it was saddening. On the other end, for essential businesses offering in-demand products, it proved difficult to keep products on shelves.

The high demand eventually resorted in businessowners enforcing limits on pasta, canned goods, toilet paper, meat, and on other popular items. Shoppers were desperate to stock up on goods for the upcoming months for fear of exposing themselves to the virus from frequent shopping trips; also, they bought in multiples, fearing food and supply stocks might run dry from the global-scale crisis interfering in import and food production.

Unfortunately, some exploited the demand by buying large sums of product and then profiting off it—from selling at outrageous prices—on Amazon, Shopify, or other shopping sites. People fixated on stocking up, sometimes resorted to paying high prices for below-average or fraudulent products. Across Canada, a number of complaints piled against extreme pricing.

In response to price gouging, Ontario's Premier, Doug Ford, warned sternly on live television he'd "go after" those hyping prices, mandating up to a $300,00 fine for banking in on the pandemic (Global News, 2020). Also, Doug Ford and Alberta's Premier, Jason Kenney, called out particular businesses, cautioning they stop after receiving several complaints, if not in the hundreds, of their extreme gouging (CTV News, 2020).

But, isn't that how business works? Actually, the involvement of the government is a part of our free market system; we're a mixed market in Canada, considering it authorizes the government to step in and intervene, when necessary. And, when considering the severity of the pandemic and Canadians' need for masks or hand sanitizer, an unfair 200% or 400% spike in prices required intervention from governance (Global News, 2020). So if you were curious, that is why officials stepped in.

In comparison, Alberta acted later than its counterparts on greedy listings, as laws hadn't previously defined when a product entered into traversed grounds; that being said, the newly implemented price-gouging COVID-19' protocol went as followed: a warning first could be issued, but if a business chose not to comply, it then potentially escalated to hefty fines (Dormer, 2020, para. 7-9). That is what officials agreed upon as the "internal policy" for COVID-19 gouging (para. 8). Outside accounts of overpricing still remain void of legislation, but the Consumer Protection Act will likely be redefined after the pandemic (para. 16).

Without the spike in pricing, the pandemic already taxed us physically, emotionally, spiritually, and, yes, financially. Leaders took consumer complaints seriously by ensuring products were free of radical pricing, as most did not possess the money to purchase such inflated products. Despite the government's financial intervention, the pandemic brought the rise and fall of businesses, while also heightening the most unlikely of products to royalty. In this next section, we will discuss the prized possessions of COVID-19, the extent people went to for them, and what we learned, as a result, from this messy pandemic. We mean, to understand why we squandered and behaved the way we did for beans and toilet paper would be interesting. What a lethal combination too, I might add.

Wiping Paper

Mm, how to convey this sticky situation. Frankly speaking, we're certain some oddballs, out of necessity, portioned-controlled toilet paper during Phase 1 of the pandemic. Could we blame them? Gossip, social media, and empty shelves had fooled Canadians into believing in a supposed toilet paper shortage, although the 250% increase in demand had "disrupted the demand, throwing it out of sync" (Proctor, 2020, para. 9-29). To also add to the shortage, because toilet paper is bulky, businesses don't usually "hold too much inventory, [as] it [usually] just flows" (para. 28).

There you have it. Toilet paper, often flowing naturally in its supply loop, was disrupted in its cycle from the high demands, making shelves skimpy and lagging (para. 28).

Cotton Masks Now a Fashion

I'm certain you remember when people were scrambling to buy masks. After Canada had given a chunk of its national stockpile to China, some provinces, namely Saskatchewan, used protective equipment from the last H1N1 outbreak, as they "were stored in optimal conditions" and passed inhalation, exhalation, and filtration tests (para. 14-16). In hopes to replenish stockpiles, Canada counted on China to produce 1.8 billion units of protective gear for the country, despite the outbreak affecting the manufacturing process (para. 7-16).

It was officials' recommendation to wear masks in public settings—combined with mask shortages and outrageous pricing—that catalyzed people to sew masks, purchase them locally or via online shopping. But, the mask is now taking on another connotation, as the heat of the pandemic has fizzled.

The cotton mask, now a quintessential fashion piece of the pandemic, is expected to evolve into an accessory for future wardrobes, post-pandemic (WCAX-3, 2020, para. 16). Although masks were seldomly worn in Western countries before COVID-19, moving forward it will likely be more commonplace and will launch as "a brand-new genre of accessories" (para. 12). Businesses and clothing retailers are trying to capitalize on the mask market, ever evolving with society and into the possible future of retail (para. 11).

A CLEAN FREAK'S DREAM OR NIGHTMARE COME TRUE: THE HAND SANITIZER SHORTAGE

Besides the eventual shortage of sanitizer, certain disinfectants have been recalled by Health Canada, as they pose health risks or lack proper labelling (Government of Canada, 2020, para. 1). While disinfectant can be convenient for using after shopping trips, some of them contained risky ingredients, such as, technical-grade ethanol or ethyl acetate, or failed to mention them in risk statements on labeling (chart 1).

While Health Canada "temporarily allowed the use of technical-grade ethanol in alcohol-based sanitizers" amid the pandemic, according to the Government of Canada, "hand sanitizers that contain unacceptable grades or denaturants that are not approved for sale in Canada have not been reviewed for safety or efficacy" (para. 4). Denaturants, which make sanitizers of a putrid smell, are added to prevent ingestion by children, but non-approved types of it have been added to some products, specifically, ethyl acetate and methanol (para. 4). After frequent use of ethyl acetate, a person may develop dry skin, irritation, or cracking. But, from methanol use it may "cause dermatitis, eye irritation, upper respiratory system irritation, and headaches" (para. 4).

The Government of Canada recommends consumers read the labelling, check for the listed recalled ingredients, report any complaints to Health Canada, and contact doctors or professionals if you have used these products and have health concerns (para. 1).

CONCLUSION

Although some overly capitalized on the pandemic or incorporated risky ingredients in products, especially, in hand sanitizers, it is our duty to look out for attempts of scamming or fishy sales. It is also apparent that in speedy production of high in-demand items, quality and pricing were sometimes sacrificed. More than anything, it reveals how quality stems from time well-spent. However, in terms of the pandemic, speed is what production required, as health practitioners and citizens needed protective gear. Despite the unexpectedness of the pandemic, we eventually had the products needed to better protect ourselves and others from the virus. A micro-digression, we have to use protective gear or products wisely and according to the recommended guidelines to obtain its maximum effectiveness though.

Largely, the pandemic has made health professionals and leaders rethink protective supplies, as the hole in our supply chain exposed itself (Leo, 2020, para. 28). Considering the usefulness or preventative nature of these products (toilet paper, masks, and hand sanitizer), it is obvious why people wanted them and sometimes resorted to paying a pretty penny. Or, we mean, a hundred dollars or two.

Chapter 8:
Misinformation in Academia

Well, well. It appears misinformation has sunk its teeth into social media, but also into the academic world, namely some academic journals, textbooks, and conferences (Smith, 2018, para. 2). Academic research, like journalism, has held a position of authority in society for decades; except, with recent journals regurgitating misguided pandemic reportage and other later-to-be-named reasons, its influence might dissolve to that of journalism, if it persists. So, as you guessed it, given that concern, we will be discussing how the misinformation crisis has permeated yet another communication medium, specifically, the academic fabric.

Wait—ahum. Before you doze off thinking about how boring of a topic academic misinformation might be, we encourage you to give it a chance; you'll likely be surprised about the behind-the-scene activities occurring. To tell you the truth, academia misinformation has been intriguing to research, and, rather Renaissance of us to say but, it induced the most adequate amounts of moustache stroking.

Background Information

To start, we must be cautious in scapegoating social media purely for academic misinformation, considering other offenders—as equally damaging—require correction, including: phony journals, fake reviewers, and out of context translations (Smith, 2018, para. 9; Ridgway, 2020, para. 9). Yes, there's more than one area in need of problem-control. To tell you the truth, it's a toss-up in pinpointing the factor most damaging, as they all work towards discrediting academic research.

From analysis, misinformation per say has not soaked the academic sleeve as deeply as social media, given the rigor of most journals and the lack of immediacy in publishing. Still, academia must persist in maintaining its sphere of influence, by not losing itself to the tug of mainstream gossip or predatory journals; it is detrimental in upholding any foreseeable credibility. And, our concern for its credibility isn't misguided either, after considering the blasts in the media against journalism. In today's world, communicators need to squeak of a clean track record to ward off jives from the "fake news" detectives swarming online. Though, that sometimes doesn't help anyways.

But, in other words, if journalism has been discredited for suspected misinformation, how will academic research bypass any repercussions for publishing inaccurate details? Then, as a result of becoming a discredited source, where will people go for information? Do guidelines need to be re-evaluated? How can we address the problem? Yes, although academia and journalism may not be the same ball game, journalism seems a quintessential example of what could evolve from malpractice in research.

After further consideration, perhaps the invincibility ascribed to research has contributed to facts running astray or overlooked at times. Academia holds an invincible rapport within society, but maybe, just maybe, it disguises the need to address poor practices. However, from headlines and some "just-leaked" conspiracy videos,

COVID-19 appears to have opened some conversations about the faults in the academic world. And, the discussion of it embodies both good and bad factors. What could possibly be bad about it?

The consistent discussion of faulty research could inspire fake news trolls, constantly hammering researchers 10 years from now, and we don't want things to transcend that far (psst… a little mental gymnastics, but we're thinking about damage control). But, then again, most importantly, discussion about bad research in the media should urge researchers to correct faulty practices or to continue in conducting quality practices.

By taking recent criticisms seriously, researchers can prevent its future rapport from shrinking to the worst kind of crumbs; by the way, the really crumbly one's too. However, if mistakes run uncorrected, that could lead to serious credibility issues, to that of journalism. Anyway, now that we've covered some depressing background information, let's zoom in on COVID-19's academic misinformation.

COVID-19's Academic Misinformation

Without a doubt, accuracy remains the hallmark of scholarly research, but the paramount importance in curing COVID-19, and fast, has left research open to error (Ridgway, 2020, para. 9). So, what contributed to the errors? Well, during our global-scale pandemic, tighter deadlines and thousands more of research articles flooded in, falling into the lap of tired, rushed reviewers (para. 4).

The urgent need to submit research findings on the virus, coupled with increased pressure, and time constraints made studies succumb to, as some suspected, not as rigorous fact checking or peer-reviews (para. 9). The notion of producing quality content, within some submitted work, drowned in the noise of our pandemic, as COVID-19 required immediate action.

It is well-known that the standard for academic journals is to factcheck and then factcheck again, as any errors can potentially discredit the researchers and publication. However, during the pandemic, a time of exhaustive supply and demand, a few questionable articles or facts went overlooked, unintentionally or intentionally (para. 4).

In previous chapters, we identified social media as being a platform, carrier, or a breeding ground for sticky misinformation. While social media still proved a hot mess for misleading news, academia contributed in its own ways to the pandemic confusion, sort of. You'll see what we mean.

In our analysis of academic misinformation, unique qualities exhibited themselves, as, oftentimes, the source of inaccuracy stemmed from often unheard of reasons. In this next section, we expand on those areas, those boards and loose nails that built the fortress of wobbly academic misinformation.

But, before we point our saucy fingers on those pressing issues, we acknowledge that while several recent papers were a smudge misguided or disproven, it alluded to how science builds upon itself. Important to realize, it is our job, as citizens, to accept new discoveries and to filter out the old (Banks, 2020, para. 22). In other words, not all articles were intentionally deceptive; they were simply in-progress masterpieces.

We might seem a tad critical of academics too, but it is to uphold the standard of excellency research has striven for and credits itself to, and, ultimately, to shine a bright light on the fake journals, deceiving the world or the research community. In this next section, you will likely find it rings familiar to Chapter 4: Phishing and Scams. Let's dive in.

PHONY SHMONY JOURNALS

Ah, yes, the phony shmony journals. We've heard the stories; it can be perilous for writers to start writing or to get their foot in the door. Sometimes novice writers sign up with all-too promising journals to only submit articles with questionable findings. While it is unknown if the writer conducted professional motives or not, or if the publisher altered facts to fit their ideological agenda, the facts somehow wobbled into misguided territory (Smith, 2018, para. 7-8). Smith (2018), a previous Criminal Defense Investigator and now librarian from Stanford University, found the following:

> The infiltrators are predatory journals and imitation conferences that have hijacked legitimate enterprises and are impersonating them for monetary gain. Predatory for-profit journals have increased at an alarming rate. In 2015, more than half a million authors had their papers published in predatory journals. It is uncertain how many were hoodwinked by sneaky publishers versus how many were willing partners with professional motives. Examples of this fake news incursion includes predatory journals with pretend peer reviewers. (para. 7-8).

As Smith discussed, some journals disguised as reputable sources for money or for some other gain. In Chapter 13, we will discuss how sometimes the intent of misinformation stemmed from an ideological gain, rather than from money. But, returning to the findings of Smith, she claimed more than a half a million writers submitted articles to "predatory journals" (para. 7-8). Does that infer further distinction between good and bad journals needs to exist? Apart from sources, sponsorships, and the date indicating an article's validity, could a stamp or a seal further distinguish journals, helping assist individuals who may be unfamiliar with the telltale signs of bad information? Or, do studies require two peer reviewing boards before publication? Do we need to slow down production deadlines and have employers emphasize quality over quantity? Or, do organizations need to provide greater support for writers, who often gain little in return, in their pursuit of sharing information to the world? There are many factors which contribute to the rise in misinformation. However, it is obvious that "predatory journals" need to be addressed at the root and labeled for what it is, given the rise in confusion and discrediting it aspires against academic research.

MISATTRIBUTION OF AUTHORSHIP

If you ever happen to be on the dark web, don't be shocked to read: "Attention: Authorship for Sale." Yes, a black-market for authorship listings exists, according to a recent 2019 study from South Korea (Rivera, p. 3).

Rivera suggested the intention behind misattribution is the following: "instead of behaving ethically, [researchers] opt for misattribution of authorship to avoid a disadvantageous position in front of academic rivals" (p. 4). Sounds awfully competitive to work in the academic world. Interesting.

In the next section, it then highlights "a man first-listed authorship" gets published more often, though results show men are more guilty of seeking misattributed authorship than women (p. 2). At first, we expected women had compromised in their research, by giving into misattribution, but that doesn't fit the data collection. Another factor, aside from the gender gap, contributed to the misattribution of authorship.

To the researcher, unethical behaviour evolved from

the overwhelming push and desire to publish (para 1). While those motives weren't wrongful and morally distasteful, how those motives were dealt with traversed into malpractice; it inspired the surfing of the dark web or paying for unfit peer reviews. This next bit reveals how to concentrate efforts, since the researcher identified geographical patterns in the cycle.

While unethical practices occurred on a worldwide-scale, including in western countries, the majority of cases concentrated in Asian countries (p. 1). That concentration arose from a combination of poverty and a growing number of low-quality periodicals, ultimately leading to ethical pursuits becoming an afterthought (p. 4). In summary, 'a lack of' resulted in compromised or inadequate peer reviews (p. 1). What can be done to relinquish research from these influencing factors?

Social Media Versus Academia

Now, onto social media and the academic world. Believe it or not, belief in science is declining in Canada, according to survey from last year in 2019. In fact, in Canada alone, the survey suggested "32% of Canadians are skeptical about science and 44% believe scientists are elitists" (Banks, 2020, para. 22).

What is stirring unbelief in science and research, aside from constant consumption of conspiracy videos or pseudo articles, may be because people altogether discredit science once a study goes disproven (para. 22). After a person doubts scientific research, it then becomes difficult to convince the deceived otherwise, or to show how a line of thought or how a "fact" may be inaccurate (para. 17).

Regardless of whether or not a person believes in academic research though, an influencing factor in pandemic behaviour, according to a study by McGill University, comes from news consumption.

According to Cardenas (2020), "people who get their news from social media are more likely to have misperceptions about COVID-19" (para. 1). And also, "those that consume more traditional news media have fewer misperceptions and are more likely to follow public health recommendations like social distancing" (para. 1).

In conclusion, rebellion to distancing may not be whether or not a person believes in science or if a person has suspected ills of research, it may overflow from their news consumption. However, if a writer recycles bad reportage or borrows information from social media for their research papers, that can lend to issues, especially when it isn't properly peer reviewed material or has later been disproven.

Lastly, in the tens of thousands of newly published research papers, a mixture of good, bad, and slightly misguided research has been spotted (Ridgway, 2020, para. 4). Some of the misguided studies reflected bad reportage, old notions, or fictitious, made-up findings (para. 4).

What to do? The following list details some necessities to safely navigate academic discussions: factchecking, opening discussions with researchers, accessing multiple sources, and second-guessing outdated findings.

Ultimately, for consumers or writers of research, upholding factchecking methods before sharing information needs to return as the centrepiece of research; it is a matter in upholding future credibility and diminishing the growth of misinformation. We need to crack down on inaccurate musings, preventing yet another "fake news-journalism" disaster.

CHAPTER 9:
HEALTH EFFECTS OF MISINFORMATION

"The world will never be the same." This is a saying that is commonly circulating over news and media outlets. COVID-19 has changed our current life and likely our future. Going to the gym, conversing with coworkers, and even getting together with friends was taken for granted by nearly the entire world until this event. How do we get back these precious memories? How can I live my life without fear of getting sick, or even worse, spreading my sickness to a loved one? While these questions may be too complex to answer, one thing is certain: we must have accurate and knowledgeable information regarding health effects from COVID-19 in order to establish some normality in our lives. Unfortunately, this is a tough task. Global leaders, world health organizations, and media outlets are often giving conflicting advice on how to handle the virus. The repercussions of this are indeed dire. I will go through several examples of misinformation regarding COVID-19 and explain how this negatively affects our health and well-being.

The greatest and most influential spreaders of information are global leaders. Presidents and Prime Ministers around the world are there to protect their people. Furthermore, they incorporate systems that promote health and prosperity. If every global leader was successful in accomplishing this feat, then indeed, our world would be more of an utopia. Unfortunately, this is far from the case. Currently, it seems every leader is on a different agenda and follows a different script. Living in Canada, the majority of the population has taken this pandemic very seriously. This is partially because our Prime Minister, Justin Trudeau, advocated for stringent stay at home measures, and to wear a mask when in public places (Coletta, 2020, para. 16). He placed responsibility on every citizen of Canada to do their part in controlling the virus. The idea is that you are not just protecting yourself from the virus, but you are protecting the vulnerable. Let's say that instead of the virus affecting more elderly people, the young and youthful were at greater risk. It's likely that nearly every grandparent or elderly individual would take extreme measures to ensure the safety of younger generations; so why don't we do the same for them? Posing these types of questions helps us understand the importance of not just taking care of our own well-being, but the well-being of our family, community, and country.

It is a proven fact that masks save lives and stop the spread of the virus (Centers for Disease Control and Prevention, 2020, para. 2). It is in place to prevent saliva and bacteria from transferring from our mouths and onto another person (World Health Organization, n.d., para. 1). According to the World Health Organization, it is highly irresponsible and selfish to neglect these facts. Luckily, Canadian health organizations and Justin Trudeau advocate for wearing a mask when in public. This cannot be said for other countries. The United States and Brazil are two countries that come to mind that completely neglect face mask usage. Trump has consistently reinstated that his country has the virus under control and has failed to publicly wear a mask when giving interviews or in public. Only after his country has reached nearly 3,300,000 cases and 135,000 deaths has he decided that it is appropriate to wear a mask (Worldometer, 2020, chart 1). Brazilian president, Jair Bolsonaro has also failed to publicly wear a mask. He has expressed that the virus is just a mild cold and that there is little to worry about (Ricard and Medeiros, 2020, p. 4) . Ironically, Bolsonaro is now recovering from the virus, and his country currently has 1,880,000 cases and 72,920 deaths (Worldometer, 2020, chart 1). It is no surprise that the two countries whose leaders failed to understand the dangers of the virus are now the two countries with the highest number of cases and deaths in the entire world.

Further, misinformation that plays down the virus is highly dangerous to everyone's health. "The virus only has a 1% mortality rate." "The virus only severely affects the elderly or those with underlying health issues." "Even if I get the virus it will only feel like a cold." Statements like these are dangerous and dangerously misleading. For every six COVID patients, one will have difficulty breathing, and, additionally, one in five will require hospital attention (World Health Organization, n.d., para. 2-22). Further, the long-term effects are still largely unknown for recovering COVID patients. Those with underlying heart and lung issues that have recovered from COVID are now at greater risk for future problems (Heart and Stroke, 2020, para. 6). According to the Advisory Board (2020), physicians have reported that patients hospitalized from COVID-19 are experiencing high rates of blood clots related to strokes, heart attacks, lung blockages, and other complications (para. 6). Physicians have seen a rise in the number of strokes in young patients due to COVID-19 (para. 2). Blood clots are also increasingly common (para. 3). Pulmonary embolisms can occur, which clots block circulation to the lungs (para. 3). Functional limitations ensue such as shortness of breath, heart palpitations and discomfort when performing physical activity (para. 3). Given these risks, it is naïve and selfish to take this virus for granted.

On April 18th, right in the heart of the coronavirus, Florida's beaches opened. Despite an executive order limiting 50 people gatherings at one time, the beaches were flooded with beachgoers (Pasquini, 2020, para. 5). This faced heavy backlash especially on social media where people expressed outrage over the decision made by GOP Governor DeSantis (para. 6). It must be noted that Florida is not the only state that faced criticism. Plenty of beaches around the world have opened prematurely (Maak, 2020, para. 17). Consequently, the vulnerable people of the world will suffer the consequences for these actions. As Maak states, "call it common sense or call it 'self-leadership', as citizens, we all have a responsibility to heed the current health advice or risk increasingly severe government restrictions" (para. 1) Do your part; be a good citizen. We all own and share the COVID-19 responsibility. It doesn't matter where we are in the world, we must act together to stop the spread.

The belief that social distancing does not flatten the curve, or the belief that there is no curve to flatten, is misinformation that greatly affects an individual's health. Without social distancing, cases would rise at an even higher exponential rate. Deaths would surely follow suit. According to Augusta Health, social distancing is not a new idea (2020, para. 6). It is over 100 years old and history has proven that it does indeed work (para. 6). In 2007, two studies published in the Proceedings of the National Academy of Sciences studied the Flu Pandemic of 1918 (para. 6). Specifically, the study analyzed the spread of the virus in different cities in the United States (para. 6). Fatality rates, times and public health interventions were all compared and analysed. The results showed how rates were roughly 50% lower in cities that implemented preventative measures, like social distancing (para. 6). This was compared to cities that implemented social distancing practices later or not at all. The cities that were most effective simultaneously closed down schools, churches, theatres, and banned public gatherings altogether (para. 6). This slowed the spread of the disease and allowed for vaccine development, as well as lessening the strain on hospitals (para. 6). Another conclusion drawn from the study was that cities that felt secure and lessened social distancing practices because of their low case levels and low death toll experienced a new outbreak. Otherwise known as a second wave. Cities that kept the intervention in place did not actually experience a second wave of high case numbers and deaths (para. 7).

Given this study and amongst others, it is proven that social distancing works. We can even analyze our contemporary world to justify this. The United States for instance laxed social distancing practices very prematurely. This has resulted in the highest case numbers and death toll in the entire world. The Centres for Disease Control and Infection has outlined ways in which we can improve our social distancing measures and keep our community safe. This information is viable, credible and trustworthy. 1) Know before you go: Before going out, understand and follow local health and safety guidelines where you live (para. 9); 2) Prepare for Transportation: When running errands or commuting to and from work, consider social distancing options (para. 11). This could be walking, bicycling, wheelchair rolling, or using public transit, taxis and ride share. Be weary when using public transit that you are maintaining a six-foot distance from other passengers and the operator (para. 14); 3) Limit contact when running errands: Only visit stores selling household essentials in person when you absolutely need to (para. 13). Drive thru services, curb side pickups, or delivery services are other viable options to limit face-to-face contact with others. It is imperative to maintain physical distance between yourself and service providers during an exchange. Facial protection should also be worn (para. 8). 4) Choose safe social activities: One is able to stay connected with friends using video chat, home calling, or through social media. If you are meeting others in person, stay at least six feet apart from others who are not from your household. Small outdoor gatherings and yard or driveway gatherings with a small group of family or friends is ideal; 5) Keep distance at events and gatherings: It is safest to avoid large gathering and crowded places (para. 30). In these areas, it may be hard to social distance and keep six feet apart. Furthermore, pay attention to physical guidelines such as tape markings on the floor and walls. This always directs attendees to remain six feet apart from each other in lines; 6) Stay distanced while being active: In a time where human beings struggle to be active, it is important to consider doing physical activities such as walks and bike rides. It is important to be aware of safe locations in your neighborhood where social distancing can be done. If you decide to visit nearby parks, trails and recreational facilities, it is important to check for closures first (para. 5).

Currently, the global medical community is fighting to find a cure for COVID-19. As of right now, no cure is available. One must be cautious when listening to advice about possible cures or remedies for COVID-19. This misinformation can be deadly. United States' President Donald Trump advocated for ingesting a malaria drug to protect against the coronavirus (Miller et al., 2020, para. 2). This drug is called hydroxychloroquine. According to Miller et al., Trump has spent weeks pushing the drug as a potential cure or prophylaxis for COVID-19 (para. 2). This was against the advice of many top health care officials in the medical community (para. 1). This drug has the potential to cause severe side effects in some patients and has not been shown to combat the coronavirus (para. 2). Last month, "the Food and Drug Administration warned health officials that the drug should not be used to treat COVID-19 outside of hospitals or research settings" (para.

6). This is due to sometimes fatal side effects (para .6). The regulators issued an alert for the drug, of which is used to treat lupus and arthritis (para. 6). Reports of heart rhythm problems, including deaths, were received from poison control centres and other health providers (para. 6). When questioned about these reports Trump hastily responded, "All I can tell you is, so far I seem to be OK" (para. 7). Trump had reportedly taken the drug for several weeks at the time of his remarks. Given his older age of seventy-three years, and increased weight gain, many consider ingesting this drug to be very dangerous for Trump (para. 8). Schumer on MSNBC commented how Trump may not actually be taking the drug given. She noted that because he lies about things so frequently, who knows what the truth is (para. 11). She also added that if he is taking the drug, it is "reckless, reckless, reckless" (para. 11). Dr. Patrice Harris, president of the American Health Association, has stated that there is no evidence that hydroxychloroquine is an effective treatment or the prevention of COVID-19 (para. 16). Further that the results to date are not promising (para. 16).

We must be cautious when given information about the coronavirus. When misinformation circulates, one's health and well-being can be jeopardized. It is essential and imperative that we only trust credible sources. Unfortunately, as described above, this is increasingly difficult because of so much conflicting information. In order to preserve the safety of our family, neighbors and community, we must find ways to encourage the accurate circulation of information. If we are to surpass this pandemic without even more devastation, we must all be on the same page.

CHAPTER 10:
SOCIAL EFFECTS OF MISINFORMATION

During this global pandemic, not only is one's health at great risk, but also a person's social well-being. Being stuck at home self-isolating, people are likely susceptible to increased anxiety levels, drug abuse, and domestic violence (Premier Health, 2020, para. 1-24). These factors are all indirectly related to COVID-19. Misinformation has also heightened these negative elements; we have a serious problem not only in Canada, but also on a global-scale. It is essential that we deeply analyze the social implications from COVID-19 so that accurate information can be readily available. Furthermore, at a time when Canada is still recovering from COVID-19, how do we know what sorts of things are socially responsible and safe? Should I go to a restaurant or not? Am I putting people at risk if I play team sports? Is it really safe to send my kid off to school? Misinformation spreads rapidly at a time when uncertainty is high. It is important to know the risks of doing particular things and the precautions one can take to mitigate this risk. In order for any country to recover as quickly as possible from this pandemic, everyone must be on the same page. We must work together to ensure credible and trustworthy information is reachable to each and every person. In this chapter, I am going to touch on how misinformation about the virus affects people on a social-level.

The origins of COVID-19 is a thorny, triggered subject worldwide. What we do know is that COVID-19 was first documented at Wuhan's Hunan Seafood Market (Campbell et al., 2020, para. 13). Wei Guixian, fifty-seven years of age, was one of the first cases to catch the novel coronavirus (para. 13). She was a woman who worked in the market every day, selling shrimp (para. 13). She had developed a fever in the middle of December, but brushed it aside as a mere seasonal flu (para. 13-14).

According to Campbell et al., unfortunately, "a week later she was drifting in and out of consciousness in a Chinese hospital ward" (para. 14). This was the start of the global pandemic, as far as we know. Evidence suggests that the virus transferred from wild animals to humans, and that bats are the likely origin for COVID-19 (para. 20). Still, there are countless possibilities. It is suggested, though, that the virus jumped from "bats to humans via a pangolin intermediary" (para. 22).

According to Campbell et al., "As many as 2.7 million of these scaly mammals have been taken from the wild across Asia and Africa for consumption mostly in China" (para. 22). There is a Chinese belief that these "scales treat everything from rheumatoid arthritis to inflammation" (para. 22). Currently, this is the most popular hypothesis for how the virus started (para. 20). However, this contradicts other beliefs. Some suggest that this virus was constructed in a Chinese laboratory or that this was a political move made by the Chinese government. As of now, no withstanding factual or information ties the Chinese government to the coronavirus (AFP Staff, 2020, para. 5). Believing in these types of conspiracies only generates the flow of misinformation. Unfortunately, the implications of doing so can create a form of social shaming against the Asian community.

A survey conducted between June 15th-18th by researchers at the "Angus Reid Institute and University of Alberta asked 516 people of Chinese ethnicity about their experience with racism during COVID-19" (Shah, 2020, para. 3). The results found that half of them indicated being called insults or names that were directly a result from COVID-19 (para. 4). A staggering 43% mentioned that they faced threats or intimidation (para. 4). One man noted that he even had his Canadian citizenship called into question despite being born in Canada (para. 6). Findings show 30% of the sample size being frequently exposed to racist graffiti or social media posts since the start of the pandemic in March (para. 10). According to Shah, "Nearly 30% said they have often been made to feel as if they are a threat to the health and safety of other people" (para. 13). Individuals have even changed their daily lives as a response to an uptick of racism (Shah, 2020). Six of ten respondents have made changes to their lives in order to prevent unpleasant situations (para. 14). 8% have even indicated facing physical abuse frequently, while 21% face physical abuse infrequently (para. 23). Kimberly Noels, a University of Alberta psychology professor, believes that "all Canadians should pay attention to the results" (para. 6). She notes how, "it's important to keep in mind how hurtful these incidents are" (para. 7).

Blaming the cause of the virus on a particular country reeks no positive outcome. United States' president Donald Trump has referred to the coronavirus as the "Chinese virus or "Wuhan virus." Using terminology such as this "personify the threat" (The Conversation, 2020, para. 9). When using the adjective "Chinese," it associates the infection with an ethnicity. This type of language flares anxiety, fear and even disgust towards that group (para. 10). We must be cognoscente of this and work to eradicate this kind of terminology from our vocabulary. Greater awareness on how people of Asian ancestry are impacted by this virus will help to decrease racist altercations from happening.

Social distancing has increased stress and anxiety levels. Businesses have shut down, families have been torn apart, and there is still no clear end in sight. Compounding on this, opioid intake and deaths have increased dramatically because of the virus (CBC, 2020, para. 1). Because of this, more information and services need to be readily available to Canadians. Depression levels have also increased dramatically, as some have increased drinking, are grieving for lost income, or are stuck at home, since the pandemic (Craggs, 2020, para. 4-10). Overdose prevention sites are still running. However, physical distancing guidelines have limited the number of people able to use these services. According to CBC News (2020), a site in "Toronto that once averaged over 100 visits a day has now seen fewer than half" and "preliminary results from the Ontario's coroner's office suggest roughly a 25% increase in overdose deaths from March to May 2020 compared to the same three months stretch last year" (para. 2-4). The statistics continue. In British Columbia, "a 39% increase in overdose deaths in April compared to the same month last year," while "the number of opioid related calls to EMS in Alberta increased from 257 in March to 550 this May" (para. 10-11). It is extremely evident that increased awareness about opioid usage needs to be a priority of the Canadian Government.

As noted previously, depression and anxiety rates have increased dramatically since the start of the virus. Undoubtedly, as mentioned in Chapter 5, this pandemic has taken a toll on the social and mental well-being of everyone affected. According to Sher (2020), "53.8% of 1210 respondents felt the psychological impact of the outbreak to be moderate or severe" (p. 2). Furthermore, "16.5% reported moderate or severe depression symptoms and 28.8% reported moderate to severe anxiety symptoms" (p. 2).

Sher also read and compared several studies, concluding that people around the world were experiencing distress, anxiety, fear of contagion, depression, and insomnia, but that healthcare workers were "especially distressed" (p. 2). Sher also concludes, "this is consistent with the results of a recent Kaiser Family Foundation survey indicating that 45% of adults in the USA report that their mental health has been negatively impacted due to worry and stress over the coronavirus" (p. 2). Moreover, in the United States, alcohol sales rose by 55% in the week ending March 21, 2020, compared to figures from the same period last year (p. 2). Online, alcohol sales increased by 243% (p. 2). There is significant research that relates increased alcohol consumption with greater depression and anxiety levels (p. 4). When trying to social distance or in an increasingly isolated environment, people have become more depressed and suicidal during the pandemic (p. 2). While alcohol can ease short-term stress and anxiety, in the long-term, if alcohol amounts increase, it can cause more problems or later turn into an addiction (para. 7). Truly, alcohol can be a harmful approach in coping with the pandemic or with other life crises, as it doesn't fix the problem, but amplifies problems through altering our body's hormonal balance (Buddy, 2020, para. 27).

According to the National Academy of Medicine, in order to reduce suicides during the COVID-19 crisis, "it is imperative to decrease anxiety, stress, loneliness and fears in the general population" (as cited in Sher, 2020, p. 4). Social media should be used as a tool to promote mental health and reduce anxiety levels (p. 4). Eating healthy foods, getting appropriate rest, maintaining relationships via phone or online, and staying active are all essential factors that will decrease one's overall anxiety (p. 4). Furthermore, as discussed in Chapter 2, setting limits on media consumption can improve our mental health and overall mood. Media includes social media, local and world news

(National Suicide Prevention Lifeline, n.d., para 3). Once again, it can help to get accurate information from reputable resources.

In Canada, a trustworthy source for health information is the Centre for Disease Control at cdc.gov. Contacting local healthcare providers or calling 211 or 311 services are also feasible options. Resource and crisis lines are available, designed to help Canadians who need help. If you are experiencing emotional distress related to COVID-19, the National Distress Line is available at 1-800-985-5990 (National Suicide Prevention Lifeline, n.d., para 3). Although the pandemic has caused challenges or restrictions on in-person support, the Lifeline Crisis Centre is active and available to the public (para. 5).

According to the Centre for Disease Control and Prevention, an important facet to further decrease stress levels is to actively stop the spread of rumours, and "[to] understand the risk to yourself and the people you care about, [as it] can encourage connecting with others, [thus] mak[ing] the outbreak less stressful" (2020, para. 8).

An often-overlooked vulnerable group who are highly susceptible to COVID-19 are the homeless population. According to the Government of Canada (2020), "structural factors, insufficient services and individual circumstances is often the result from homelessness" (para. 2). Those experiencing homelessness may pose at a higher risk of contracting COVID-19 or might undergo further complications, as there are breaches in obtaining regular services and resources during the pandemic (Government of Canada, 2020, para. 3).

It is reported by Aboriginal women and girls, along with other gender diverse people, that it is increasingly difficult to access shelters at this time (para. 3). Inherently, officials fear this may hinder their ability to effectively quarantine, self-isolate, physical distance or exercise hand hygiene (para. 3). Other complications the homeless people may face may involve their higher rates of asthma, chronic obstructive pulmonary disease (COPD), and heart conditions (para. 3). Further, the homeless populace, on average, also has more medical conditions related to their physical health, mental health and substance usage (para. 3).

Mental health support should also be made available and emphasized during this fragile time. There simply isn't enough information regarding the protection of this vulnerable group in our society. In response to the pandemic, as expected, shelters have decreased in capacity to adhere to pandemic safety measures. This, as a result, leaves many to face increased hardships in dealing with substance abuse and other health disorders without the proper support.

Domestic violence cases have also steadily increased since the start of the pandemic. There has been a 20-30 percent increase in gender-based violence across the country since the start of the pandemic (Patel, 2020, para. 1). According to Monsef, in some places calls for help increased by 400% (as cited in Patel, 2020, para. 4). Meanwhile, the York Regional Police recorded "a 22% increase in domestic incidents since stay at home orders came into effect March 17" (para. 7). However, as emphasized by Monsef, fewer calls does not equate to fewer domestic violence' incidents (para. 8). Instead, victims may be failing to report incidents, as they are being closely surveillanced (para. 10).

In response to the rise, the federal government has delegated $50 million to women's shelters, sexual assault centres, and to similar organizations within the Indigenous communities to assist those during the crisis (para. 12).

The funds are to help facilities adhere to social distancing measures, thus continuing operations during the pandemic (para. 13).

McGinnis, from the Wheatland Crisis Society, noticed how "numbers were down in women shelters, partly because of reduced capacity" (para. 14). However, McGinnis also believes numbers are down because people are unaware of the current extra measures implemented in the home shelters to harness against COVID-19 (para. 14). Originally McGinnis' centre had received 333 crisis calls, but in March, after social distancing measures, the number dropped to 203 phone calls (para. 17).

McGinnis notes how, "A pandemic doesn't make (violence) stop. A pandemic just makes that silent" (para. 18). Lisa Martin, executive director of Women's Shelters Canada, explains how the Canadian government needs to communicate how a person doesn't need to stay at home if it isn't safe (para. 24). It is clear that not enough information and attention surrounding vulnerable groups is available or is, potentially, the product of misinformation.

The social effects of misinformation can be extremely detrimental to our society as a whole. From prejudicing an entire ethnicity, to not vocalizing the problems faced by marginalized groups, one's social health has faced many barriers during this period. It is essential to spread accurate information to those who are vulnerable and need help in a time such as this. We are all global citizens of this earth, let's all do our part for each other.

Chapter 11:
Psychological Effects of Misinformation

Overall mental health worldwide has been on the decline ever since COVID-19 globally started in February 2020. Our whole world has shifted: people have been asked to stay at home and not visit their loved ones, there have been shortages at the grocery stores, and major income losses; all of these have taken a massive toll on psychological well-being. The spread of misinformation sadly only intensifies this issue (Centre for Disease Control and Prevention, 2020, para. 8). While the current COVID-19' pandemic continues to be a prevalent issue in countries across the world, causing physical illness, it also seems to have a huge impact on mental and psychological health worldwide on individuals and communities.

In fact, people who have never contracted the virus or who have come in contact with someone who had the virus are reporting "significant psychiatric morbidities, negative emotions, and poor psychosocial and coping responses toward the outbreak of infectious diseases and consistent worry about contracting the disease" (Mukhtar, 2020, p. 512).

During the first few months of the spread of the coronavirus, information, news updates, stories and conspiracy theories were flooding the news and the internet. It was almost impossible to avoid seeing something relating to the virus every time you switched on your local news channel or every time you were browsing through social media. In addition, with the stay at home protocols, people had just that much more time to spend watching the news, or browsing through the internet, which exponentially increased exposure to the novel virus's "infodemic." It was seemingly hard to escape this realm of panic spreading through the media and social networking sites. After a while seeing these things over and over, day and day out really can have a toll on people's mental health and psychological beings.

In addition, this plethora of information and news updates being circulated, there has been a huge outflow of misinformation going around the internet, which has drastic effects on people's mental wellbeing (Centre for Disease Control and Prevention, 2020, para. 8). To make matters even worse, the more time people spend at home, the more time they have not only to view all this misinformation, but to also spread or share it with their family, friends, or social media followers, who, in return could do the same.

This misinformation has led to racism, uncertainty, and anxiety revolving contradicting information. During a study, "nearly 58% of people who spent more time on the social media platform Reddit reported that the overall psychological being was decreasing, while only 22% of people said that reading about COVID-19 online found their mental health to be better" (Suciu, 2020, para. 11). On the other hand, people who have been making efforts to limit their social media time and their time spent exposing themselves to COVID-19 coverage were "more likely to report improved mental health" (Suciu, 2020, para. 5).

Although with social media being such a prevalent resource in our society's everyday lives, avoiding it or "turning off the screen" is easier said than done, not to mention only more difficult with stay-at-home measures being taken.

No one should have to fear for their mental or physical well-being while walking down the street, going to pick up groceries, going to work, or while doing other daily activities. No one should have to fear for their child's well-being, after sending them to school. A big issue due to misinformation, surrounding the coronavirus, that is affecting the mental well-being of the Asian community involves the uprising of blame, racism and bullying towards people of an Asian background.

As mentioned in earlier chapters, the World Health Organization believes that the COVID-19 virus began in Wuhan, China, but then spread worldwide. Even in Canada, many Chinese-Canadians have reported name-calling, being intimidated publicly, threatened, or even physically attacked by a stranger. In a survey study from the Angus Reid Institute, in partnership with the University of Alberta, they uncovered some shocking realities from over 500 Canadians of Chinese ethnicity (2020, para. 4). The researchers found the following:

> *Half (50%) report being called names or insulted as a direct result of the COVID-19 outbreak, and a plurality (43%) further say they've been threatened or intimidated. Additionally, three-in-ten (30%) report being frequently exposed to racist graffiti or messaging on social media since the pandemic began, while just as many (29%) say they have frequently been made to feel as though they posed a threat to the health and safety of others. (para. 5-6)*

This level of racial discrimination can have a detrimental impact on the mental state of Asian ethnicities, specifically depression and anxiety. It has even come to the extent that some of the people who participated in the study above have reported that they are afraid to leave their house or have "adjusted their routines in order to avoid run-ins or otherwise unpleasant encounters since the COVID-19 outbreak began" (para. 8). These micro-aggressions toward the Asian community have left them feeling isolated and have overall declined their overall psychological state. Some have reported even feeling anxious sending their children to school in fear that they will be bullied (para. 10). These increased anxieties have shifted their everyday lives. Not to mention that these added stresses are on top of all of the anxieties that the rest of the population are facing due to the virus.

Furthermore, a large number of Asian owning businesses have been suffering; some report customers suddenly acting strangely or seeming afraid to support their businesses. A flower shop owner in Toronto reported customers were questioning "if it was safe to approach him anymore due to his Asian background, and asked if he had ever been to Wuhan, China" (Gill, 2020, para. 4). Unfortunately, the rise of misinformation surrounding the coronavirus has led to an uprise in racism and microaggressions toward the Asian community, leaving their mental health in a suffering state.

Another issue that is affecting people is the overexposure to a ton of different information, which oftentimes gets pretty contradictory. It seems as if you can find an article online supporting any belief you may have in regard to the coronavirus, whether it be to wear a mask or not, stay home or not, and so forth. People who are already panicking and who are already anxious will grab onto any sort of "credible" information they can find, and later find a contradicting message that same day.

This uncertainty and confusion in the people definitely lead to heightened anxieties, panic, and a lower psychological well-being (Sher, 2020, p. 2). Misinformation is not only damaging psychically. For example, remember the couple who consumed the chloroquine in their fish bowl cleaner and ended up in the hospital (Sheperd, 2020, para. 4-5)? But, yes, it can also take a serious toll on mental health as well. Oftentimes, fake news is created to invoke an emotional response from its viewers, and can invoke feelings of "anger, suspicion, anxiety, and even depression by distorting our thinking" (Erdelyi, 2019, para. 16). Furthermore, this information then gets passed along online or through word-of-mouth, either intentionally or unintentionally; it also poses the risk of getting even more distorted, as it gets passed along from person to person; especially, in cases where the said information is being translated from language to language, which is happening a lot in this case as the pandemic is affecting countries from across the globe. This can cause an increase in panic and fear and can potentially lead to more serious psychological issues. In fact, even when an article is debunked as fake news, it still can create negative emotions within the population as they often can be left feeling frustrated or angry and, in some cases, powerless against this plethora of misinformation (Margit, 2020, para. 16).

 In saying that, people really need to be smart about where they are receiving their information from, especially when it comes to serious health matters, such as a global pandemic. Misinformation can have a serious impact on people's health whether they recognize it as fake news or not. It is recommended to evaluate all information that is presented to us and to fact check before we accept it as the truth.

Another issue that affects our mental health due to the misinformation that is circulating online is that it is creating a divide between the people, further than simply physical. While people are stuck at home already separated from loved ones, friends, coworkers, and peers, they are now being further divided with contradicting beliefs about the virus; some of which stems from misinformation being spread or shared online or in-person. People are feeling more isolated than ever, which has a huge effect on one's psychological being, and can even lead to depression or anxiety.

For example, we will look at the whole wearing a mask situation. It has been recommended by health officials that masks should be worn when in public or when not able to social distance, and some cities and states have already made it mandatory in hopes to curb the spread of the coronavirus any further. According to Mandavilli (2020), researchers "have estimated that about 40 percent in the general population might be able to be infected without showing signs of it" (para. 26). In saying that wearing a mask is simply a sign of respect since even though seemingly healthy people are able to contract and spread the virus without even knowing it (Garcia de Jesus et al., 2020, para. 5).

Although it seems rather simple to cover your nose and mouth while going out in public, people on and off social media have a lot to say about the matter. There have been several disputes over wearing masks or not, and some even led to physical assault or even death (Dawson et al., 2020, para. 1). People on social media have twisted the narrative of wearing a mask to a political statement. They are claiming that they refuse to let the government control them so in that case they refuse to wear a mask that could protect them and those around them. If you and one of your friend's family members or peers were on opposing ends of this or a similar controversy, it could lead to psychological distress, strain on relationships, even further isolation, and risks resulting in even further mental health decline.

All in all, the coronavirus and the facts surrounding it are causing enough of a strain on the general population's psychological well-being, and adding in the effects of all of the misinformation spreading around does nothing but make it even worse. People are worrying about being separated from their family and friends, unemployment, financial issues, their children's education, and also the fear of contracting the virus, but now they are left worrying about a plethora of fake news either reaching them or their loved ones. All of this stress compiled could have detrimental effects.

Unfortunately, mental health is not being focused on enough by medical professionals or on the news, as everyone is putting more effort into focusing on the amount of cases or deaths of the virus. People's psychological states are just as important as their physical health and need to be addressed and treated appropriately. As we discussed in the last chapter, several studies suggested that as the coronavirus started spreading, people's mental health began to decline. But also, factors such as racism and bullying, and contradicting articles have resulted in confusion or in division of the people, which could lead to some serious mental health issues. People need to make sure they are taking not only physical precautions but also mental ones, ensuring they are of wellstanding psychological health. People also need to make sure to reach out to medical professionals in case of depression, anxiety, or other mental illnesses arise.

Chapter 12:
Risk Factors of Misinformation

Misinformation has been around for a while, but more so in recent years; since the 2016 face-off between American political candidates, Donald Trump and Hillary Clinton, misinformation has centered political discussions, news reporting, and social media discourse. While it is not the discourse of misinformation that is alarming, though one might argue otherwise, it is the growing number of misinformation-consumers that's raising flags. And not to raise anyone's blood pressure or trigger any panic, but misinformation, our loud friend, is in need of boundaries, as it has surpassed toxic-Mordor-rumor-mills and workplace gossip, seeping into social media and YouTube channels—a place the world connects on multiple times a day.

It is likely, at this point, you've already encountered many forms of misinformation in the media, the bug slipping into the nooks and crannies of political debates and music lyrics. It is a bug growing in size and impact, after all. But, what about misinformation do we possibly not know? Let's explore.

To accentuate the scale of misinformation, in the last three months of the 2016 American election, professionals recorded the 20 most-popular fake election stories roused

1.3 million more shares, like, and comments than the 20-most popular, legitimized election stories online (Hambrick & Marquardt, 2018, para. 1).

Since then, sources have also credited misinformation to influencing election polls, while other researchers disagree, claiming fake news goes detected more than suggested or feared (Shalby, 2019, para. 4). But, aside from whether or not it is influencing elections, let's look at misinformation' mishaps on a COVID-19 basis.

More recently, and as a by-product of the COVID-19 pandemic, 700 Iranians died from methanol consumption, thinking ingestion caused immunity to the hard-hitting virus (Aljazeera, 2020, para. 1). Iranians being unaware of the characteristics of alcohol, as drinking is prohibited in the country, led to consumption of the methanol beverages. Some sources say alcoholic Iranians, buying their regular fix of alcohol from bootleggers, died from the methanol sold to them, rather than from their efforts to kill the fast-spreading virus; while, several other witnesses say they drank "alcohol" to war against the virus because of promising social media posts (ABC News, 2020, para. 8; Malekian, 2020, para. 4).

Without a doubt, and as the facts testify to, misinformation contributed to the sum of deaths. From February 20th to April 6th, 2020, when the first wave passed across borders, 728 Iranians had reportedly died from alcohol poisoning, while in the last year only 66 alcohol-related deaths were ever recorded (Aljazeera, 2020, para. 2). What else could account for the increase in alcohol poisoning and for the increase in injuries and deaths during this particular time frame?

What do we collect from the situation? Some "educational" YouTube videos or bogus social media posts may appear promising, harmless, or endearing, but, as we witnessed with the Iranians, they can inflict serious harm, including not only extreme polarization or blindness, but the death of friends and family members (ABC News, 2020, para. 7). During any crisis, it is detrimental to question social media posts, to ask professionals about cures, and to research proposed beliefs posted online. Unfortunately, as we discussed in Chapter 4: Phishing and Scams, a crisis urges criminals to exploit resources for personal gain—and that is what likely happened in Iran and abroad.

Apart from this depressing situation fueled by misinformation—the death of several Iranians—there have been several other areas of risk identified by professionals. In the next few sections, we will then discuss how professionals have so far tackled misinformation, the main sources of risk during the pandemic, and some preventative measures.

WHAT PROFESSIONALS ARE DOING FOR PREVENTION?

In the meantime, and as of now, our warring against misinformation, which Merriam-Webster Dictionary defines as "incorrect or misleading information", entails education about media usage (p. 7764). Whether education has been effective or not is up to debate, but, we strongly believe, access and awareness of fact checking resources is the next step and most crucial in debunking misinformation.

While resources are available to the public, it is of sneaking suspicion that few use them or know how to access them (Scheufele & Krause, 2019, p. 7764). Considering that, user-friendly fact checking tools may be a means to look into and integrate into media channels, accessing them whilst scanning our feed. It is likely of value to integrate consumer-education into high school curriculum too, as it then reaches not a select few, but an entire generation.

Crackdown on COVID-19 misinformation, during the pandemic, has been constant as not only does misinformation sometimes lead to dangerous actions, but it discredits popular news outlets, furthering the ensnarement of conspiracy videos or "real" fake news (Tandoc et. al., 2017, p. 1). COVID-19, largely, has revealed the harms of misinformation as people, across the globe, died drinking bleach or by consuming large doses of chloroquine medicine (The Department of Global Communications, 2020, para. 4).

While some forms of misinformation appears harmless, deceptive information can grow to a monster, producing unkindly attitudes or unnecessary deaths. In the next few sections, we will then discuss some of the risk factors of misinformation during the COVID-19 pandemic and the cost of slippery misinformation.

Risk Factor #1-Contiguity of the Virus from Anti-Mask Culture

Several misinformation videos and posts rebel against the use of masks. This anti-mask culture emerging in communications poses a risk as it threatens the safety of ourselves and others.

We know. Masks can be discomforting and sweaty, but studies have shown the spread of contagions minimizes from distancing and wearing a 12 to 16 layer cotton mask (Adigun & Johnson, 2020, para. 5).

But, what about personal choice? While, yes, you have a choice, of course, depending on the location or the circumstance, a mask is advised or encouraged to be worn in public, especially in crowded or bustling places. Any initial questions about mask-wearing is no longer a topic of debate as they were resolved at the end of May (para. 2). What ended the debate about whether a mask was effective or not was from a test done in China by The Lancet, revealing people should wear a mask and distance 3-6 metres from each other (para. 1). They concluded, mask-wearing lowered transmission and distancing safeguarded against the flow of the virus, ultimately preventing further repercussions and unnecessary harm.

Ultimately, to wear a mask is to help one another and protect our neighbors. It is harder, sometimes uncomfortable to, but, as a rule of thumb, it is always better to run on the side of caution than to inflict unnecessary suffering.

RISK FACTOR 2- THE CREDITING OF FAKE NEWS AS REAL NEWS

Social platforms are often used to blast journalists for misinformation. The discrediting of authentic news sources only furthers the spread of misleading news, giving grounds to a vicious cycle (Tandoc & Ling, 2017, p. 12).

 Mass confusion over which news outlet to trust sometimes leads people elsewhere for news updates, largely to social media (p. 3). However, these unofficial news sources fail to be informative or accurate, but successfully appease the views or speculations of a targeted audience (p. 13).

Unquestionably, discrediting the news and resorting instead to misleading "reportage" can sway people to break social distancing or to polarize in beliefs, causing division. Change in news outlets does not necessarily mean a person will rebel against COVID-19 guidelines, but the transition to low quality, misinformed content can inspire concrete actions (p. 1).

This is where the conflict and sparks fly. While freedom

of speech is a given right, and fully supported, when misinformation places others at risk, it crosses into a grey area. Does that grey area, a sticky landmine trapping murky information potent with risks, require intervention, as it no longer becomes about constructive news or critical ideas, but transcends into being a lethal, harmful voice of misinformation? See, the conflict arises from the bounds of freedom of speech. Is it plausible to intervene or to block a video or two? Can we trust the bigger guys with censorship not to cross boundaries? The series of unknowns kindles friction.

To add to the discussion, researchers suggest the motivation behind "fake news" reportage is financial and ideological gain (p. 2). More clicks and shares equates not only more revenue, but influence (p. 2). According to Tandoc and Ling (2017), "when a post is accompanied by many likes, shares, or comments, it is more likely to receive attention by others, and therefore more likely to be further liked, shared, or commented on" (p. 3). Popularity on social media likens to a self-fulfilling cycle, as it "lends well to the propagation of unverified information" (p. 3). Meaning, popularity spreads misleading posts further, as they circulate to a larger audience and then onto a larger audience and so forth.

It is alarming because misinformation has the power to influence others into false ideas, reaping a wave of questionable beliefs or havoc in their lives or in those around them. While we do not encourage readers to accept information unquestionably, we encourage all our readers to be critical thinkers about all the information we consume and uphold in our lives. In hindsight, it really is a saddening thought to live a life based on fears or lies.

Risk 3-What Happened Again? The Misinformation Effect

An old theory, called the misinformation effect, from the 90's is being discussed again in psychology periodicals. According to Cherry (2020), the misinformation effect "refers to the tendency for post-event information to interfere with the memory of the original event" (para. 1). In the studies, researchers found that introducing new information, after an event, changed a person's memory of an event or situation (para. 2).

While the misinformation effect pertained to concerns over eyewitness accounts in criminal cases, the theory also shed insight into the workings of misleading information. It showed how although a person may intend to tell the truth, proposed or outside information distorted our memory, forming false ones (para. 1).

Influencing factors, believed to form these false memories, include the following: time, the discussion of events with other witnesses, news reports, and repeated exposure to misinformation (para. 14-17).

To expand on repeated exposure, the researchers suggested the "more a person is exposed to misleading information, the more likely they will believe the misinformation was a part of the original event" (para. 17).

If the researchers' findings are, in fact, correct, that calls into question our own memories and what others may believe about the virus because of misinformation. A scary afterthought, but exposure does have an effect on our memories and thoughts. One way to protect against false memories, researchers recommend to immediately write down what occurred during an event on paper (para. 18). It leaves room to only subtle errors (para. 19).

Final Thoughts

Final thoughts, interactions with social media need sifting to that of something more cautious and in-tune with reality. Social media isn't any less different than meeting new people, as we are, in fact, engaging with new people and their ideas, except online. In other words, the way we approach social media is in need of some tuning. Misinformation is likened to a fire ignited from a gust of wind. Once its blown, it travels far, depending on the direction pulled by the wind. We, in this case, though, can control the wind and the spread of false information by not sharing knowingly false information online; it is a small, but big contribution to the world. Remember the old adage that goes: "I tried to change others to make a difference, but then I discovered it started by fact checking my social media post."

CHAPTER 13:
EDUCATION AND WAYS TO PREVENT MISINFORMATION

Misinformation and fake news are terms being thrown around in the political ring and amongst Canadians lately. The terms' meanings couldn't be more of a question mark, though. In a basic Google search of "define misinformation" or "fake news," a glut of ambiguous terms and indefinite explanations appear, only furthering confusion and the misdeeds of misinformation.

A major contributor to the heaps and piles of buzz words and fake news lingo was no other than in 2016, during the presidential showdown between Trump and Clinton in the United States (Egelhofer & Lecheler, 2020, p. 97). Words, besides misinformation and fake news, that also buzzed into popularity since the campaign, include none other than the highly abused—fake news, gate keeper, echo chamber, disinformation, and on and so forth. Each term points to a specific type of misinformation, ranging from propaganda to tunnel vision binoculars, aka the constant consumption of one viewpoint. Since the presidential election, misinformation buzz words have been integrated in rebuttals by scientists and politicians alike and have exploded on social media pages in the comment sections (p. 97).

To clarify the meaning of misinformation though, according to Merriam-Webster, it is "incorrect or misleading information" (n.d., para. 1). A simple definition for a big-scaled problem, but it is still a comprehensive explanation. The World Health Organization also re-introduced the term "infodemic" during the COVID-19 pandemic. Although some credited WHO with the term, Merriam-Webster suggests the word was coined in 2003 from a Washington Post column by David Rothkopf (n.d., para. 2).

If you are wondering, infodemic is basically a fusion of "information" and "pandemic," and "refers to a rapid and far-reaching spread of both accurate and inaccurate information about something, such as a disease" (para. 1). As we can already see, the burst of information inspired by the pandemic not only shook cobwebs off of old misinformation lingo, but became iridescent in its design, being incorporated more uniquely and richly in daily communication over time.

MISINFORMATION RISKS

Some of the communications entertained during the pandemic became dangerous at times, when taken into practice. During COVID-19's boom of inaccurate, and sometimes flat dangerous, information, Google, Facebook, Pinterest, TikTok, Youtube, Tencent, and other platforms took efforts to filter content that encouraged questionable practices, such as, to consume high doses of chloroquine medicine, incorporate loads of ginger and garlic in drinks or food, or to stay outdoors in hot weather, as the virus "fizzled" to that of nonexistent in high temperatures (The Department of Global Communications, 2020, para. 4).

I know, I know. Upon reading, the flag on garlic may come off as bizarre, as that is a stinky hallmark of cooking often measured by the heart and known for its health benefits—including, reducing total cholesterol or defending against colds—however, garlic was inapt in stopping the transmission of the virus. Reports had shown that garlic demonstrated zero effect on warding against the coronavirus (Web MD, 2020, para. 2). Frankly speaking, garlic cannot replace social distancing or quarantining, though, jokingly, it may inspire it. Media platforms, and a decent chunk of them, had monitored these headlines and keywords to protect the public from potentially placing themselves or others in compromising situations. Again, the actions mentioned above are some of the unsuitable methods of protection against the novel virus.

COMPARATIVE OF MISINFORMATION HISTORY AND NOW

Misinformation from the past infers a specific kind of content, not the kind of Tony-bologna-fake-news spoken of today. Mountains far from today's meaning of misinformation, past misinformation originally meant news parodies, political satires, and news propaganda (Tandoc & Ling, 2017, p. 2). News parodies and political satires, and I would even argue propaganda, had lumped into the misinformation category, considering intentions were not-so obvious at times and open to misinterpretation by the public (p. 7).

With rise in social media use, we see misinformation definitions mirroring online discourse. Definitions also hint to the concern for accuracy in news outlets. The drift from reading newspapers or watching broadcasts on television, people now get their news on social media, but that poses a problem in combating misinformation; it only escalates and repeats the suspicious and misinformation cycle, as social media often goes unchecked (p. 3).

In return, these definitions further pry at the authenticity of news outlets or inspire people to sniff for fishy facts, suspected of being misrepresented or of swaying others to thinking akin to the news outlet or to their "agenda" (p. 12). This next observation blows the head off the gasket, inspiring food for thought. According Egelhofer and Lecheler (2020), "it is when audiences mistake it as real news that fake news is able to play with journalism's legitimacy . . . Another dimension of this is that fake news needs the nourishment of troubled times in order to take root" (p. 12-13). How relevant. During the pandemic, these two elements described were seen in action: 1) multiple voices commented on the virus, leading others to question news sources, and 2) society was in a troubling, uncertain time, searching for answers. Troubling times, unfortunately, only ripen with misinformation, especially when multiple voices are commenting at large. In this chapter, we will then address sources of misinformation and how to fact check content, so we can be beacons of stewarded information online.

TYPES OF MISINFORMATION

There are two streams of misinformation that all misleading content falls under: misinformation and disinformation. The distinction between the streams, misinformation and disinformation, is intention. As we discussed earlier, misinformation is "information that is false, but the person who is disseminating it believes that it is true" (Wardle & Derakhshan, 2018, p. 44). Now, please pay special attention to the motivation in the next definition. Disinformation, the other side of the spectrum, is "information that is false, and the person disseminating it knows it is false. It is a deliberate, intentional lie, and points to people being actively disinformed by malicious actions" (p. 44). In the condensed SparkNotes version, misinformation means a person spreads phony information, thinking it is true; while, those sharing disinformation, know it to be false, but deliberately share the information out of wrongful ambitions (p. 44).

Now that we have discussed the two motives behind sharing false information, let's look at the types, the categories, of bogus content flooding online channels. In each section, consider how it may fall into disinformation or misinformation, and what authors obtain by using this tactic. Also, consider if it is appropriate or not, and what the ramifications may look like.

SATIRE AND PARODY

Oftentimes, this humorous genre leaves more heads scratched or outraged than any other category.

Fueled with witty, over-the-top critiques, satire can easily be misunderstood by the audience (Tandoc & Ling, 2017, p. 5). Depending on the execution, satire can be more informative about political affairs than official deliveries, but it is "acknowledged to have significantly shaped public discourse, opinions, and political trust" (p. 6).

Parody shares various similarities with satirical content, as they both use comedy to draw in an audience, but facts usually poof into thin air in the parodical genre (p. 6). A concern for parodies, like for satirical pieces, is they be interpreted as an official news source, rather than as a spin on events (p. 6).

FALSE CONNECTION

Often false connections are used in clickbait headlines. False connections happen "when headlines, visuals, or captions do not support the content (Wardle & Derakhshan,

2018, p. 47). Not only do many people find clicking on clickbait articles frustrating or disappointing, but oftentimes misleading. The risk of only reading a headline or the beginning of an article can lead to false conclusions, the spread of faulty information, and irrational reactions from readers. The aim of these articles are to get clicks and to entertain, regardless of whether or not a reader feels deceived.

FALSE CONTEXT

Taking a situation and applying it sometimes brings another level of understanding, but it only works when the context fits. False context is "when genuine content is re-circulated out of its original context" (p. 47). It can be thought provoking to come across this information, whether by visuals or through writing, but upon referring back to the source, to only see it distorted or misapplied . . . how misleading. From what I've noticed, experts in a subject or those adamant on researching the original content weed out this deception, as they know the intended context.

IMPOSTER CONTENT

Imposter content is when a disingenuous source pretends to be an official site, news outlet, or organization. Upon an organization learning of a risky imposter account or article floating about, often they will alert followers to discredit the article, breaking off any affiliation. Imposter content can be cunning as imposters use an author's byline or a company's logo in videos or images (p. 47).

MISLEADING CONTENT

Watch out for claims used to blame or target an individual. Claims, and even assumptions, can be oh-so misleading. Misleading content is using "information to frame an issue or an individual" (Cameron University, n.d., para. 4).

Framing can be less discrete too, as it can be "by cropping photos or choosing quotes or statistics selectively" (Wardle & Derakhshan, 2018, p. 47). This misleading device is also known as the "Framing Theory" (p. 47). A way to sniff one out can be in analyzing if the argument is balanced or in-favor of one position. Also, asking the subject for a response is indispensable as they can refute claims at the root or supplement our understanding, leading to other discussions or more digging.

MANIPULATED CONTENT

Remember the Putin and Trump meeting that sparked attention, as the picture detailed others closing in, in their serious discussion? What were they talking about? Well, sadly, no one remembers since Putin was never at the meeting and only photoshopped into the alien-looking hot seat (Sheth, 2017, para. 3-4).

As said in the name, manipulated content is when content, which can be photos or information, is intentionally altered to deceive (Cameron University, n.d. para. 4). Sensational as the Putin and Trump image was, unfortunately, there was only so much to say about the imaginary meeting, but that it was a sham.

FABRICATED CONTENT

I think everyone can hear Dwight Schrute from The Office in the background at this point. False.

Fabricated content is completely and utterly false. Fake content often can be found in stories, images, or websites (Bellemare, 2019, para. 21). The websites or outlets often appear as legitimate sites that have a small following (para. 27).

Some fabricated sites will pose as a local newspaper or

Toronto-based periodical, but are actually located in a faraway country, such as, Ukraine or elsewhere (para. 28). The hosts of the website aim to gain revenue from online users through subscriptions or advertised merchandise (para. 28). Upon further investigation, a person may discover the newspaper ceases to exist and is only a money grab.

FACT CHECKING

An awareness of the types of misinformation not only helps protect us from harmful information, but it refines our perspective on subject matter. The internet is a swamp of good and bad information, but it is our onus to factcheck and to critically engage with content. More than ever, it is important to critically engage with the following list, when consuming media online:

1. Scan for fishy or questionable sources—old, not credible?
2. Evaluate how it made you feel—did it spur a reaction or rage?
3. Consult other articles and see what they say—look for inconsistencies and consistencies in stated facts.
4. Look at listed sponsors—has the sponsorship possibly painted their words or perspective?
5. Check dates—is it past its expiration date or recent?
6. Examine, critically dissect the ideas presented
7. Apply common sense
8. Ask a professional
9. Use a fact checker tool or run a reverse image check on Google (other tools work too)
10. Come to terms with our own biases, and again
11. Be honest with our own biases (Mind Tools Content Team, n.d., p. 1).

Applying a critical lense to our information consumption

not only serves to uncover a new idea or refined perspective, but it gives us back the control we crave while surfing the web. It can lead to finding answers and bring us to informed opinions.

Fact checking can be likened to dating or meeting new people. We have to ask questions, pay attention to what is communicated, and then act accordingly. Content is not all quality content. Quality people exist, but also do people not acting in quality.

CHAPTER 14:

COMMUNICATION AND THE SPREAD OF INFORMATION AFTER THE PANDEMIC AND IN THE FUTURE

Back in February of 2020, WHO said COVID-19 is "more than just a pandemic–it is an infodemic" (D'Amore, 2020, para. 2). Much like a disease, maladaptive information can be nebulous, omnipresent, and infectious. The problem of misinformation is not a new phenomenon, it has existed for as long as humans have been able to communicate. What makes it so salient is precisely why this pandemic is so devastating: humans exist in a globalized world. Our supply-chains, our travel plans, our families, our information, all flow across varying nation-states, each touched by the culture and politics of each region. Information itself is uniquely incorporeal, you cannot track it as easily as one can follow pineapples from Hawaii or iPhones from China. It isn't necessarily a product that carries receipts but an exchange that transpires through various vectors: a newspaper, a bulletin, a friend, to name a few. However, this generation's flavour of globalism and interconnection is defined primarily by the worldwide web. It is by this medium that an unspeakable volume of information is flooded into the news feeds of attentive consumers.

With the rise of social media giants like Google, Facebook, and Twitter, the curation of information for specific tastes and preferences has never been easier. Such catering has brought about innumerable advantages for people, but it also has come with an increasingly apparent cost. While people can more readily connect with others who share their preferences, this innocent fact swiftly distorts into social myopia as the slippery slope of preferences pools people into their respective bubbles.

Communities in isolation of each other have for some time now been multiplying, diverging, isolating, and, in some cases, warring because of how the internet and social media work to encourage the reification of ideas. When ideas are a matter of preference, say, pineapple versus pizza, it is a relatively innocuous thing for random global citizens to be sparring about the matter. But when ideas are a matter of safety, where clear and consistent communication is vital, where voices that are most vulnerable can finally secure a platform to be heard, this is where danger arises. This is how infodemics are born.

What the internet and social media accomplish so handily is the disruption and redistribution of power. When the world was void of the internet, more trust was placed in the governments, universities, institutions, and corporations; they largely contributed to the narrative that the public ingested on various issues (Pew Research Center, 2019, para. 2). The internet has since shattered that arrangement. Such normative narratives required the belief of a norm, and, by extension, a belief in the abnormal (Grimes, 2016, p. 1).

For instance, it has long been established since the time of Galileo that the earth is a sphere. Though, there exists a contingent of people who identify as 'flat-earthers' that argue that the earth is not only flat, but that it is a conspiracy, orchestrated by the world's governments and space-fairing institutions, to convince the people of the world that the earth is indeed flat.

Besides the dubious amount of assumptions that would have to be verified to make that theory true, there are already numerous papers that explore how statistically improbable it is for such conspiracies to be sustained without someone tattling on these crafty institutions (p. 1-2).

Despite all of this and the breadth of resources one can easily access online to dispute these conspiracies, not only do "flat-earthers" exist, but they have grown in number. Apply this phenomenon to the present-day pandemic that, as of the time of this writing, has infected over 18 million of the world's population, locked down countries, and has confirmed killing at a rate higher than a conventional cold or flu virus at 3-4%, compared to the seasonal influenza at 0.1% (D'Amore, 2020, chart 1; World Health Organization, 2020, para. 10).

Hundreds of companies and institutions are working on vaccines, while the primary interventions remain as the following: social-distancing, wearing masks and other protective gear, increased hand-washing, and persistent cleaning of areas, surfaces, or objects in frequent contact with people (Centers for Disease Control and Prevention, 2020, para. 1-8). Without these interventions, it is foreseeable that countless others would either be infected or have lost their lives. There exists a sizable force of people out in the world, Canada included, that dispute some if not all of these realities. Realities that have been repeated with excess by scientists, politicians, and experts—those who we are supposed to trust. Yet, to the detriment of millions, so many in the world do not.

What our interactions on the internet, which manifest into our actions in everyday life, bring into clarity is our relationship with truth and authority (Tandoc & Ling, 2017, p. 1). Long have people challenged information presented by institutions with purported esteem and authority. What the internet enables though is a medium where those who reject well-established realities can isolate themselves as the majority in their own internet community. Communities can reject and accept people through manifold means, whether by blocking someone on social media to constructing fringe sites that share links on trusted forums. This creates a milieu where communities can reject conventional epistemological methodologies used by most conventional societal bodies and construct their own theories and 'alternative facts' based on assumptions they aver are true. Institutions, experts, and citizens alike attempt to combat these groups with evidence and facts that, by nature of the assumptions of many of these groups, are already rendered null. There are several reasons for this, but of primary concern is the tendency of these groups to wield confirmation biases when considering contradictory information. Show flat-earther pictures of other planets and they will likely respond by saying that the pictures are falsified propaganda touted by the very institutions designed to deceive the world's people. Show a COVID-19 denier the COVID death rates and infection rates of the globe's population and they may retort the stats are contrived, that governments are lying, or that people are dying of other causes and that COVID is just being assigned blame as a means of controlling the people and keeping them indoors.

What emerges from all this is a pattern of people challenging the mechanisms and structures that generations of curious minds have struggled to build. Without a shared agreement of how to structure, observe, and critique information, institutions cannot hope to penetrate the thick shells of communities that effectively rebuke these means. This is concerning as these structures and mechanisms that institutions speak to have been humankind's general attempts to sidestep our innate subjectivity and biases in favour of more objective information. With information untainted by special interests or cultural proclivities, the idea is that the information is more valuable to the global population and presents a truer understanding of phenomena in the world. This is especially important for widespread crises like the COVID-19 pandemic as swift, data-driven action has been observed to reduce rates of infection and deaths due to the virus (Hutchinson, 2020, para. 33). This also has the implicit effect of allowing nations to restore their battered economies much more quickly.

The threat of misinformation generated by these groups is amplified on the internet as the increased exposure makes such information seem legitimate, more akin to simply having a difference of opinion than a complete denial of widely regarded facts (Cherry, 2020, para. 14-17). This encourages average citizens, the vast majority who are not experts on issues such as pandemics, to consider themselves equal voices at the table, discussing how to interpret and approach these issues. Because of how the internet disrupts power relations, such groups are able to generate communities that appeal to the average person's sense of intelligence and control, that they can design a different path than the perceivably inconvenient solutions that established authorities provide. Many of these groups rail against the sacrifices and consequences that have emerged because of COVID-19' interventions.

With the slowing of Canada's economy and the physical restrictions in place, Canadians were forced into precarious economic situations. This, coupled with the strain of maintaining isolation, has created a wide-spread burnout (Ranosa, 2020, para. 1). Despite the onslaught of misinformation, most Canadians still trust information from institutional sources (D'Amore, 2020, para. 16). However, there still remains a sizable number that fall prey to the promises of denial-oriented online groups that preach immediate gratification and a return to normalcy to spite the perceived overreach of global institutions. This effect has been far more eminent within Canada's southern neighbour, the United States. Unfortunately, due to the physical and cultural colocation of both countries, Canada is far more prone to American media and ideologies than other places across the world. Despite many in America supporting COVID-19 interventions, there exists a sizable portion of the population that aggressively refuse to comply with regulations that restrict them in any fashion (Hutchinson, 2020, para. 4). Among other factors, America's case is unique in that their federal administration has actually engaged and appealed to denial and conspiracy oriented groups for political gain, thus, institutional authority in America has been granted to those pundits of misinformation in this pandemic (para. 25). Though, after consideration, it could also stem from repeated exposure to conspiracy videos, forming belief in speculative notions.

The net appraisal by many of these groups and of the information they espouse is that it poses an imminent threat to the health and safety of Canadians. This is why the Canadian federal government has allocated three million to combat the spread of misinformation (D'Amore, 2020, para. 32). Further, as has been evidenced in the burgeoning Black Lives Matter (BLM) movement, those who are systematically disadvantaged are more prone than people with privilege to not only contracting but dying due to

COVID-19 infections. In Canada's case, Aboriginal people are most vulnerable in this pandemic. For instance, one of the most touted of interventions is to wash one's hands. For many in Aboriginal communities, access to the clean water to do so isn't even possible (Carling & Mankani, 2020, para. 3). Further, they are far more impacted by perturbations in the economy than are any other group.

This push to combat misinformation has led many to begin calling out the catalysts for the efficacy and power of today's misinformation surge – Google, Facebook, and Twitter. For years, these companies have been charged with providing secure spaces for fringe groups, hate groups, and misinformation to run with reckless abandon. These companies have responded to such criticisms by trying to explore monitoring efforts and to expand their policies about what constitutes a violation of use regarding their respective platforms (Vincent, 2020, para. 1-6). Much of their efforts, according to critics, have been lukewarm or not readily enforced. This is also a product of how huge these companies and their mediums are. It's difficult to, in a timely fashion, follow-up with breaches in violation and other users flagging down problematic posts, accounts, and groups. Sometimes the policies these social media' giants implement are abused and their tools are used to silence others who a user or group simply disagrees with. There have even been instances when fringe groups retaliate against those who call them out by trying to get their accounts shut down with copyright claims, reports, flag–whatever the means of problem signalling are on the respective platform.

Recently, Twitter took a highly publicized step and began adding fact-check warnings and removing tweets from powerful purveyors of misinformation, like the U.S.'s President Trump (Ortutay & Hazell, 2020, para. 1). Long have these companies demonstrated unease in clamping down on misinformation from powerful sources as they wrangle with balancing the right for people to see what these powerful societal entities are communicating, while also trying not to bolster or support harmful perspectives or viewpoints. Twitter's competitor, Facebook, on the other hand, took a different approach. It's CEO, Mark Zuckerberg, remains adamant about not taking down a message written by Donald Trump that, in response to the looting accompanying BLM protests in the U.S., said "when the looting starts, the shooting starts" (Rodriguez, 2020, para. 1). This phrase originated during the 1967 race protests to expand and defend the rights of people of colour, particularly black people, in the United States (Eubanks, 2020, para. 1); Trump later stated he was unaware of the expression's meaning (para. 8). In Twitter's case, it added fact-check warnings to tweets that directly expressed such falsehoods as mail-in-voting being fraudulent, which is dangerous not only in not respecting social distancing but also to democracy; whereas Facebook's case was not one of misinformation necessarily, it was nonetheless a corporation's attempt to mitigate harm arising from a social media post (Ortutay & Hazell, 2020, para. 1; Rodriguez, 2020, para. 4).

It is undeniable that this pandemic has and will continue to change the world radically. It is difficult to predict the resonant technological and social implications of the virus on Canada post the pandemic, but the controversies that have reared during this time period may grant insights as to what we can expect to see moving forward. The rise of communities around the globe that balk at institutional information of any kind or facts that are inconvenient for their interests will face a reckoning.

Ultimately, it is the corporations dominating the social media market that will bend to the outrage of people, who are fed up with patience and the accommodations being granted to the online agents that seek to generate lies and distortions. With the damage of unrestricted communication and information posting becoming increasingly apparent, corporations will begin to follow Twitter's lead in directly shutting down these agents (Rodriguez, 2020, para. 4). Corporate arguments that weigh free speech over responsible speech will begin to crumble when it comes to blatant examples of miscommunication. This period of time has produced a multitude of traceable examples of damaging free speech and one need not look far in any journal or newspaper to find examples of the price Canada and the world pays when people are encouraged to follow their own opinions on matters of global crises: Citizens attacking store workers for being asked to wear masks; citizens flooding beaches devoid of masks or social distancing; politicians actively denying the virus' lethality or even its very existence; the list goes on (Maak, 2020, para. 17; Hutchinson, 2020, para. 10).

Given the correlation between misinformation and the behaviour of citizens, corporations, ideally with an online following, will have little choice but to expand efforts in not only monitoring, but in actively shutting down accounts that defy community policies in communications. In the meantime, users on these social media sites may become more empowered and vigorous in flagging down erring accounts. Eventually, it is likely that this will emerge as a pattern of internet activism. Companies will also begin to hire more people into roles specialized in teasing out users that violate their terms of use specifications around informational accuracy.

In order for companies to execute shutdowns or warnings regarding violations of policies around accurate information, they will first have to endeavour in defining what specifically qualifies as miscommunication. Deciphering opinion from fact, misleading information from outright lying–these will become battlegrounds that activists, legal communities, governments, and corporations will spend a great deal more time hashing out.

The communities that ultimately find themselves collapsing in more aggressive monitoring and shutdowns by social media corporations will not simply disappear. Likely, they will recede back to the nascent online forums and sites that housed them initially. However, given the popularity accrued by agents of misinformation, there will be, and even have been, apps and websites developed to specifically host and churn out the distorted realities that these agents deign to pander, whether by their own principles or ends. Parler is one such app that has been lauded by alt-right partisans as a clear alternative to Twitter (Lerman, 2020, para. 5). This gained popularity once Twitter began actively flagging and taking down Donald Trump's tweets about COVID-19 (para. 1). The app itself exists as a means for those who preach misinformation about the virus among other things to feel legitimized, to exist in a community that verifies the laws of the world that most appeals to their sensibilities. Again, the communities are most defined by their repudiation of objective analysis for their own subjective, gut instincts about the world about them. The future will see more of these apps and mediums crop up and gain popularity.

The result of corporate action against misinformation will be a further sequestration of communities and perspectives. Those who identify with specific realities will seek the mediums that justify them. A core difference from this present era will be that societal institutions that preach for more objectivity will be emboldened by the support of corporations working in tandem with them–a shared agenda to combat misinformation. Misinformation is not only compromising to our social fabric, it's compromising the bottom-line of powerful corporations. Even corporations not bound to the media are being pressured to make statements about COVID-19 and Aboriginal and Black lives; it has business implications regarding who chooses to purchase their products or services (Cassidy, 2020, para. 9). This will have the impact of undermining what has been the swelling influence and power of groups that peddle misinformation. No longer will it be seen as impressive to "think differently" from the norm if the norm is being slowly recognized, at least in respects to health and safety, as the more legitimate, just option.

Communications between people, online or not, will have this backdrop in play. Specifically, the downfall of strong individualistic, instinctual perspectives. It may not matter as greatly what stats or information you can quote as much as where the information came from. Who one follows or references may become more important to the average Canadian than just how much information one can quote on a subject. This will not necessarily bridge differences between opposing subjective versus objective positions, but it will likely reduce the seductiveness of arguments and facts built on appealing to people's innate biases and predilections.

Unfortunately, as observed in regard to BLM, when a group perceives a loss of privilege or power, they may grow more feverish in their mission to overrule trusted institutions and sources. There likely will be more protests, revolts, and violence as a consequence of agents of miscommunication losing public favour, or at the very least, seeing their opposition swell in strength. It may become risky for people to be open about their opinions on conventional topics. Even in today's world, as indicated earlier, civilians are being attacked by others in the community for even doing such things as simply wearing a mask or asking others to do so (Hutchinson, 2020, para. 11-22). The Canadian government wields enough clout and support from its citizens about COVID-19 guidelines, according to surveys, that they should be able to respond swiftly should such incidents begin to rise (D'Amore, 2020, para. 16). Laws may become more defined regarding protests' rights, conduct in public, and how law enforcement and/or social services are to mitigate or respond to such instances.

The spread of information will continue to be dominated by social media giants, even should there be a rise of alternative sources. What will be interesting to see is if a company that provides false or misleading information will rise to such heights as Facebook or Twitter in mitigating a social environment complicit to discourses that are appealing to such people. It is unlikely. In the future, successful corporations, particularly in social media, will abide by a newfound understanding: it isn't profitable any longer to appeal to the ignorant.

Conclusion

The COVID-19 pandemic has crossed borders across the globe, ever-separating and infecting people. By late September of 2020, a total of 9,217 Canadians had reportedly died from the coronavirus, since its deadly introduction (Google News, 2020, chart 1). And, numbers only continued to soar during September. In Italy, as of late September, at least 35,707 Italians had reportedly died (chart 1). In the United States, 958, 314 Americans died (chart 1). In India, 86,752 people died (chart 1). From Brazil, 136,532 died (chart 1). To accentuate the scale even more so, the mentioned countries only account for 5 of the 195 countries in the world. Worldwide 30,859,069 total cases were reported, while 958,314 deaths were confirmed as of September 20th, 2020 (chart 1). Regardless of which nation we came from, too many people have died, and we grieve for them.

Up to date, researchers are still looking for a treatment or vaccine for COVID-19. According to the World Health Organization (2020), "they are working in collaboration with scientists, business, and global health organizations through the ACT Accelerator, the joining of several organizations, to speed up the pandemic response" (para. 4). By these powerhouse organizations uniting forces, they hope the pandemic will end more quickly through "supporting the development and equitable distribution of the tests, treatments, and vaccines [needed] to reduce mortality and severe disease" (World Health Organization, 2020, para. 2).

It is only a matter of time before a cure will be found and made available. The World Health Organization estimates a vaccine will be ready for distribution sometime in summer or fall of 2021, though optimistic politician, Donald Trump, speculates it might be perfected by the end of 2020 (Radcliffe, 2020, para. 135; Facher, 2020, para. 1). Additionally, the connection between the virus and a vitamin D deficiency is currently being studied, as there may be a correlation (Cousins, 2020, para. 1). A major warning though, further research is required before researchers can confirm definitely that vitamin D-levels interplays (para. 2).

But still, while we wait patiently for the long-awaited cure, we can proceed in observing safety measures to protect ourselves and others from transmitting the coronavirus (Centers for Disease Control and Prevention, 2020, para. 1).

If you feel fuzzy on the guidelines or desire more thorough information about measures or want updates on the virus, it can be found on the Government of Canada or the Centers for Disease Control and Prevention website. A mouthful to say, we know.

If it helps, remember the MHD acronym that stands for 1) wearing a mask in public settings, 2) washing your hands, and 3) social distancing from others.

Implementing the COVID-19 safety measures, we know it will protect not only ourselves, but, most importantly, those around us who may be more susceptible or who have underlying health conditions. Those older, who constitute as over the age of 60, or who have underlying health conditions, including, diabetes, cardiovascular disease, chronic respiratory disease, or cancer, are believed to be at a higher risk, though any age group can become infected with COVID-19 (World Health Organization, 2020, para. 5).

In the months onward, it may grow tedious, unbearably sticky, or lonesome adhering to safety measures, but today's actions helps in ebbing the flow of the virus. Researchers know, as it has been concluded from countless studies, that safety measures lower transmission rates, as they minimize the transfer of droplets from person to person (Harvard Health Publishing, 2020, para. 6).

And, while we are physically distancing or following measures, please remember to FaceTime call a loved one or stay in-touch with friends during this time. As discussed, we know communication via online isn't as enriching, but it provides some interaction that can help us anchor through. There are also various hotlines and shelters available that can assist. Although initially there was a lack of vacancy, given the funding from the government, there are not only implemented safety measures in place, but more room for people and also opportunities for face to face meetings, if needed.

Additionally, as we sometimes lose focus of, please remember we won't always wear masks or socially distance, as this is only temporary. That day of freedom will soon arrive, if we stick to safety guidelines. Without any wavering, following safety guidelines ensures less loss and the virus being ridded of more quickly.

Thank you to those who wore a mask and continued observing safety measures. We know your observances in everyday living saved countless lives.

But before we let you go, we've compiled main takeaways from every chapter to refresh our memory. Here's what we've collected from the data:

Chapter 1-Studying pandemic history can diminish suspicion for safety guidelines, as our current practices stem from past pandemics.

Chapter 2-"80%-90% of people spend a whopping 24 hours a week consuming news and entertainment" (Hall & Li, 2020, para. 2). Truly, limiting our intake of media and factchecking sources, as suggested by researchers, better protects our physical health and mental well-being.

Chapter 3-Several categories of misinformation were spread on the internet and over social media; they often focused on causes of COVID-19, what precautions to take, at home remedies, and the virus being a hoax. When it is said and done, believing false information could worsen the virus itself.

Chapter 4-Access resources online to learn about scams and phishing, ultimately protecting yourself and others from COVID-19 related cyberattacks.

Chapter 5-Although business via online lowers costs, it costs us enriching human connections, dearly needed for our mental health (Wickham et al., 2014, p. 1). Does the benefits of saving money through online communications override social well-being?

Chapter 6- Leaders need the support of the science community to communicate a clear, direct message, thus evaporating the spread of wonky information.

Chapter 7-A hole in our supply chain revealed itself to health practitioners, as they lacked emergency protective supplies. On another note, please take a breath in and rethink purchases; there's some price gouging occurring in our provinces—even report them, if deemed excessive.

Chapter 8-An influencing factor in pandemic behaviour, whether good or bad, stemmed from news consumption (Cardenas, 2020, para. 1). We also need to crackdown on phony journals and academic research before it causes a trolling uproar.

Chapter 9-Misinformation can jeopardize our health and well-being. To preserve the safety of our family, neighbours, and community, we must circulate accurate information and also protect the vulnerable by adhering to safety measures.

Chapter 10-It is crucial to spread accurate information to those vulnerable or to assist by notifying people of the resources available, as it decreases mental distress (Centre for Disease Control and Prevention, 2020, para. 8). The social effects of misinformation and constant negative reportage can be detrimental to our society as a whole, considering drug use, depression, and anxiety have increased; people are also likely more suicidal, as it is a pattern concluded from past pandemics via surveys (Sher, 2020, p. 2-4). We need each other.

Chapter 11- Racism, bullying, and confusing articles have contributed to a decline in mental health (Angus Reid Institute, 2020, para. 5-6; Margit, 2020, para. 16). We need to take not only physical precautions, but also mental health precautions during the pandemic. Our psychological health is just as important as our physical health, if not more.

Remember, those limiting social media usage reported an improved mental health (Suciu, 2020, para. 5).

Chapter 12-Harmful misinformation can mature into a monster, producing unkindly attitudes or even unnecessary deaths. Ultimately, misinformation points to the need for fact checking, as the following risks amped up during the pandemic: anti-mask culture, self-medicating, further crediting of fake news as real news, and resistance to safety protocols. Misinformation truly has the potential of lengthening the duration of the virus.

Chapter 13-Whatever the type of misinformation, whether intentional or unintentional, it packs a punch, and it can lead us astray in our assumptions and opinions. Applying a critical lense to presented information and fact checking sources gives us back the control we crave while online. Fact checking and contemplating even lend to finding answers, new ideas, or even informing our opinions. Cliché, we know, but remember not all content is quality content.

Chapter 14-Social media is arming up, or at least trying to, against misinformation. In the future, will an emphasis be placed on not the ability to quote stats or facts, but specifically on the source used? It is one thing to quote a stat, but does it have social media, the great influencer, supporting its message as well? Some people are resorting to apps aligned with their views and ideologies, namely Parler, to express their ideas, but, again, the bounds of freedom of speech is an issue, as inappropriate content continues to creep up. Fortunately, major platforms are evaluating reports and policies now!

CONCLUDING POINTS

Throughout these chapters, we hope facets about the coronavirus were discovered and that the dangers of misinformation became apparent. We also hope, largely, that our readers obtained solace and feel more confident about pandemic discussions. As we often noted, the pandemic has been riddled with misinformation, which, as a result, only intensified the confusion and pain of citizens. In the future, we hope researchers and social media gurus find a way to unclog the muck of misinformation from communication platforms.

We largely believe fact checking and not sharing posts of knowingly false information can tamper the spread of misinformation. It can be comforting to watch conspiracy videos or to indulge in questionable theories, but, in the long run, it can cause strain on our mental health and in our relationships, if it influences our behaviour, negatively.

In a world swarming with messages and online updates, we hope quality information falls into your hands. Though, as that is an unlikely feat and awfully wishful, we hope you apply the tools provided and think twice about notions communicated. We know education and fact checking are important steps in prevention.

Our Salute

Together, the world grieves over the memories our loved ones gave us and for how they passed to COVID-19. We have lost several people along the way, whether it be scientists, bus drivers, or friendly neighbours not far from us. There will always be a hole in our hearts for those who passed to the coronavirus.

We are also most appreciative of those who worked the front-lines of healthcare and who still continue working endlessly to free our world from the lethal coronavirus. These selfless acts performed will forever be remembered in the ages to come.

And, thank you to those who made this book possible. A special thank you to the Antarctic Institute of Canada, TakingIT Global, the Government of Canada, and the Canada Service Corps. There is power in numbers and when people come together.

www.ingramcontent.com/pod-product-compliance
Lightning Source LLC
Chambersburg PA
CBHW071748270326
41928CB00013B/2834